D1733930

GREEN ENERGY FOR ALL

How hydrogen and electricity carry our future

GREEN ENERGY FOR ALL

How hydrogen and electricity carry our future

Ad van Wijk, Els van der Roest, Jos Boere

Foreword

Because the climate is changing, we want to switch from fossil to renewable energy. This switch involves major consequences. Not only will we soon be using other sources of energy—solar and wind instead of oil, coal and gas—the transportation and storage of energy will also have to change.

In addition, we need to look for other sources of carbon as a commodity for industry in general, and the chemical industry in particular. This book explains what is needed for a new, sustainable energy system and how it differs from the current fossil system.

In summer, there is more sun than in winter, and winds are stronger at sea than on land. So the amount of sustainable energy is not constant everywhere and all the time. In addition, solar panels and wind turbines take up a lot of space. So sustainable energy is best produced in places with a lot of sun or wind, and plenty of space. Deserts and oceans, for example. Then this energy can be transported in the form of hydrogen to where we live and work. This gas molecule is formed by electrolysis of water and it has a major advantage: it is cheap to transport and store.

Deserts and oceans are sources of not only clean energy but also of materials, water and food. For example, you can extract lithium and potassium, among other things, from brine left over after the desalination of seawater. If you make the seawater pipeline a little bigger, you can even produce water for irrigation and food in the desert. Surprising solutions—such as for the sargassum plague—are also possible in the oceans. This seaweed washes ashore in large quantities and pollutes coasts and beaches. If you capture it while it is still in the ocean, you solve a waste problem and, at the same time, tap into a source of clean energy, carbon and hydrogen.

Switching to renewable energy means more than replacing fossil energy. The entire energy system needs to be restructured. In energy scenarios, then, you have to think about variations in space and time, and the costs of transport and storage. You also have to take into account the scarcity of space in densely populated and agricultural areas because it drives up the cost of renewable energy significantly. And finally, you have to think about the use of materials because that will determine the costs of the energy system in the future.

So you can also read this book as a call to look at energy systems in broader terms: as part of an integrated system for supplying renewable energy, materials, water and food. And for a balanced assessment of a system of this kind, we introduce a set of sustainable energy system goals: clean, affordable, reliable, circular, energy supply secure, material secure, safe and fair.

We would like to conclude with an optimistic message: green energy for all is feasible and affordable thanks to the clean energy carriers hydrogen and electricity.

Ad van Wijk

INDEX

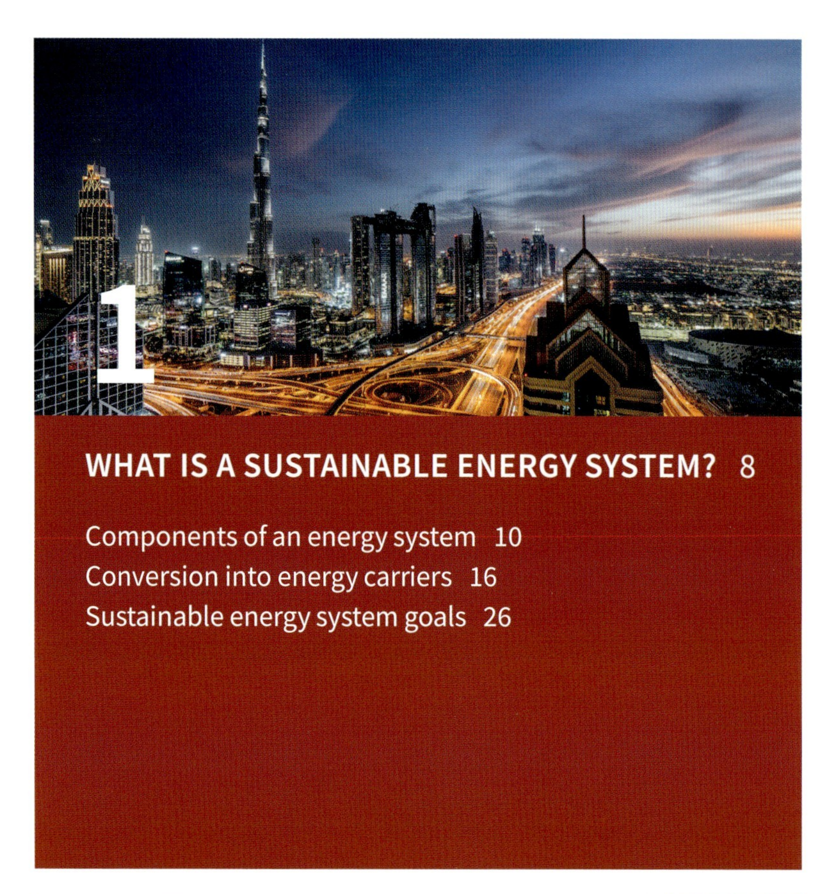

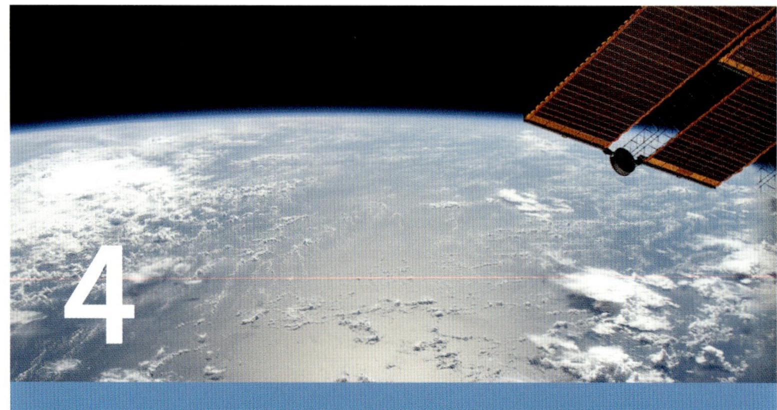

2

3

5

6

1

WHAT IS A SUSTAINABLE ENERGY SYSTEM?

Modern societies cannot do without energy. They need a well-functioning energy system that can provide energy services, such as heating or cooling your home, transporting you to and from work, and charging your cell phone. But what do we actually mean by a sustainable energy system? How do we convert sustainable energy sources into suitable energy carriers, such as electricity, heat, and hydrogen? And what are the goals of such an energy system?

Components of an energy system

As a consumer you are not always aware of it, but many energy services (such as heating and cooling your home, or charging your phone) require an energy system. Such a system can store and transport, convert and deliver energy, so that everyone has access to it anytime, anywhere. What are the components of such a system and how does it work?

What does an energy system consist of?

The goal of an energy system is the provision of energy services, for example, a comfortable temperature at home, the preparation of a meal, or the production of paper or fertilizer. Depending on the service to be provided, you need an energy carrier, such as wood, gas, oil, electricity, or hydrogen, and equipment or a system that provides the energy service. For example, if you burn the wood energy carrier in a stove at home, this provides a warm house (the energy service). The consumer of the energy service concerned receives the energy carrier via an energy infrastructure. In other words: between the supplier and the consumer of energy services there is an energy infrastructure, which ensures that the desired energy, of the right quality, in the required amount, and at the right time reaches the end-user. In order to achieve this, you need to be able to temporarily store energy in the energy system, on different timescales: from seconds, minutes, hours, and days, through to seasons. And for this, you will have first to extract the energy from an energy source and convert it into an energy carrier.

Let's take a closer look here at a few key concepts and components: energy services, energy infrastructure, energy sources, and energy carriers.

Energy services

There are five categories of energy services:

1. Heating and/or cooling of buildings.
2. Mobility (driving, sailing, flying).
3. Power and light.
4. High-temperature heat for industrial processes.
5. Energy as a feedstock, particularly for chemical products (the molecules with which products are made, such as plastics).

All of these energy services require energy, in varying amounts, depending on environmental factors, behavioral factors, specific characteristics, and the efficiency of the equipment and installations.

With the exception of the environmental factors, all these factors and characteristics can be influenced by the person who purchases the energy service. The energy required is ultimately supplied by energy carriers, each with its own conversion technology and properties.

With the specific energy use, you can easily compare the energy efficiency of two energy services with each other

Two examples of energy services, factors that influence energy use, and possible energy carriers

		Energy service	
		A warm house	**Car drive from A to B**
Factors that influence energy use	**Environmental factors**	• Outdoor temperature • Wind speed • Irradiation	• Wind speed and direction • Outside temperature • Road characteristics (e.g., resistance)
	Behavioral factors	• Indoor temperature • Thermostat settings	• Driving style and speed • Inside temperature
	Specific characteristics	• Size of home (volume in m^3) • Type of home (apartment, detached, etc.) • Insulation level of floors, walls, roof, windows	• Weight of car • Aerodynamics • Equipment in car (heating/cooling, sound, communication, security)
	Efficiency of equipment and installations	• Heat supply system • Pipes • Conversion energy carrier to heat	• Transmission to wheels • Recovery brake-energy • Conversion energy carrier to motion and electricity
Possible energy carriers		• Wood, coal, oil, natural gas, hot water, electricity, hydrogen	• Gasoline, diesel, ethanol, natural gas, electricity, hydrogen

Besides the total energy use for an energy service, the specific energy use is also an interesting factor. In the case of the heating of houses this involves energy use per m^2, while for car driving it is energy use per km. To formulate it precisely, the specific energy use is "the energy use per physical unit of the delivered energy service, with standardized environmental and behavioral factors". With this specific energy use, you can easily compare the energy efficiency of two energy services with each other.
"How much electricity does your car use? Mine does 5 (kilometers) on 1 (kWh)."

Energy infrastructure, transport, and storage

The supply of energy takes place through an energy infrastructure. Each energy carrier has an own infrastructure, with its own storage characteristics. You can however generally outline what an energy infrastructure with storage facilities looks like. A large-scale transport network is fed from large energy sources with conversion into the relevant energy carrier, and delivers directly to large-scale users, such as industry, and to the medium-sized network. This medium-sized transport network is also fed by medium-sized energy sources with conversion into the energy carrier, and delivers to small and medium-sized companies and large buildings, and to the distribution network. This distribution network is also fed by small energy sources with conversion, and delivers in turn to small energy users, such as homes. This entire delivery chain also works in reverse: the distribution network can supply the medium-sized network, that in turn can supply the large-scale transport network.

Electricity does not consist of molecules but of electrons. You cannot store them in a tank and save them for a long period

If an energy carrier is made of molecules (as in the case of wood, gasoline, gas, hydrogen, and hot/cold water), then there is the possibility of integrating (small, medium-sized, and large) storage facilities into the energy infrastructure. Short-term storage is mostly done using storage tanks, while long-term storage involves underground reservoirs, such as salt caverns, empty gas fields, or aquifers, from which you can deliver the required volumes of an energy carrier whenever desired.

The situation is different for electricity, which does not consist of molecules but of electrons. You cannot store electrons in a tank and save them for later. Small-scale, short-term storage of electricity is possible with batteries. And by pumping water up into elevated reservoirs (from which you can feed it later into a hydro-power plant), you can indirectly store a somewhat greater amount of electricity for a little while longer. But an essential feature of an electricity system is, and remains, that supply and demand need to be matched at every moment. Large-scale storage of energy for a longer period therefore always takes the form of molecules, which you can convert into electricity whenever desired. Currently, when we speak of molecules, we still mean coal, wood, natural gas, and crude oil; in the future we will also be referring to hydrogen.

Conversion of energy sources into energy carriers

Energy sources can be subdivided into fossil sources (coal, oil, and gas), uranium, and sustainable sources (solar, wind, hydro power, biomass, wave energy, geothermal energy, tidal energy, etc.). Because uranium and the sustainable energy sources are not themselves energy carriers, they first need to be converted before being used [1]. Thus, you can convert solar energy into electricity using photovoltaic cells, or into hydrogen using photolysis cells, or into hot water using solar collectors; you can convert biomass into heat, biogas, electricity, or hydrogen; you can convert geothermal energy into hot water or, in the case of high-temperature deep geothermal energy, into electricity. You can then deliver this usable energy to its destination via a transport network.

Cell phones don't work on natural gas and cars don't run on crude oil. This is why fossil energy sources also need to be converted for use. Moreover, oil pumped out from the ground varies widely in terms of its composition (of hydrocarbon molecules) and its impurities (including sulfur). Crude oil is therefore processed in refineries to produce common oil products, such as gasoline, diesel, and aviation fuel. Because these energy carriers meet uniform specifications, they can be transported and distributed through an energy infrastructure. Similarly, after it is extracted from the ground, natural gas is processed to remove impurities and produce a uniform quality. This results in different energy carriers: high-calorific gas (with little nitrogen) and low-calorific gas (with a higher percentage of nitrogen), each with its own transport system. A high-calorific gas-fired boiler does not work with a low-calorific gas, and vice versa.

As in the case of natural gas, there are also different qualities of electricity. Solar panels produce direct current (DC), but the distribution network has alternating current (AC), with a frequency of 50 to 60 Hz and a voltage of 110 to 230 volts [2]. To introduce direct current into the electric grid, you first need to transform it into the appropriate

GREEN ENERGY FOR ALL

How hydrogen and electricity carry our future

Charging an electric car's battery

alternating current. But today almost all of our appliances work on direct current. In this regard, Europe has mandated that all electronic devices must have a standardized USB port. When you charge your cell phone, the charging connector thus ensures that the (high voltage) AC delivered by the electric grid is transformed into (low voltage) DC. Since batteries also work on direct current, the alternating current delivered by the grid also needs to be transformed into direct current, before you can charge your electric car.

What does a sustainable hydrogen system look like?

Earlier we outlined what an energy system looks like in general terms. Here, we look more closely at a sustainable hydrogen system—one of the key sustainable energy systems of the future.

The energy services that hydrogen can provide, are: heating and cooling of buildings, mobility (driving, sailing, and flying), high-temperature heat for industrial processes, and electricity. You can also use hydrogen as a reducing

The components of a sustainable hydrogen system

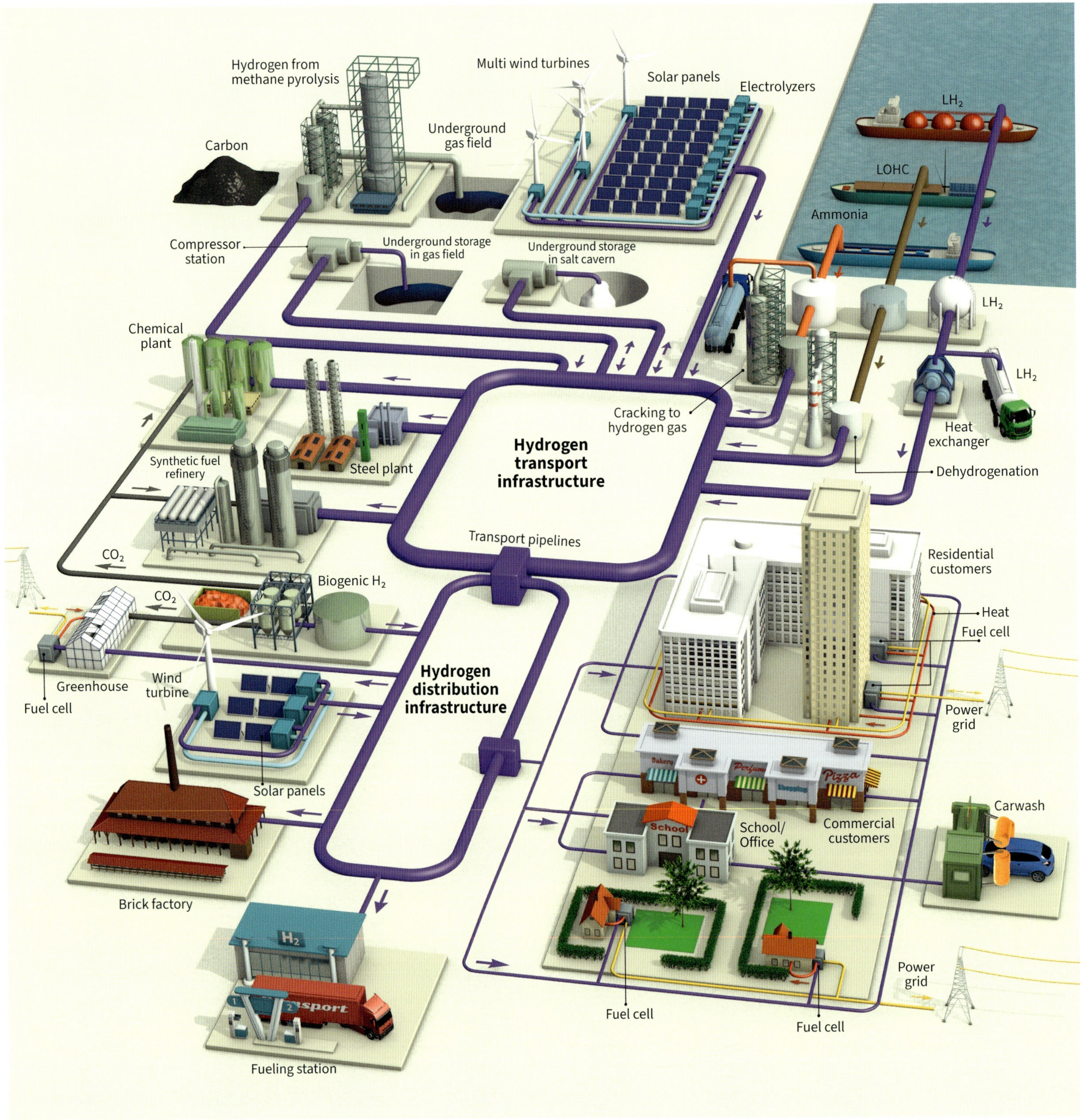

Hydrogen from methane pyrolysis
Multi wind turbines
Solar panels
Electrolyzers
LH_2
Carbon
Underground gas field
LOHC
Ammonia
Compressor station
Underground storage in gas field
Underground storage in salt cavern
LH_2
Chemical plant
LH_2
Cracking to hydrogen gas
Heat exchanger
Synthetic fuel refinery
Steel plant
Hydrogen transport infrastructure
Dehydrogenation
CO_2
Transport pipelines
Residential customers
Biogenic H_2
CO_2
Heat
Fuel cell
Greenhouse
Wind turbine
Hydrogen distribution infrastructure
Fuel cell
Power grid
Solar panels
Carwash
School/ Office
Commercial customers
Brick factory
Power grid
Fuel cell
Fuel cell
Fueling station

High-voltage grid switching station

agent for iron production, and as a feedstock for chemical products and synthetic fuels. The hydrogen infrastructure is similar to the one for natural gas. Pipelines are used for the transport and distribution of hydrogen gas. Hydrogen can also be transported by ship, but this requires that it first be transformed into liquid. This liquefaction can be done by making hydrogen liquid, by binding it to nitrogen to produce ammonia, or by binding it to a liquid organic hydrogen carrier (LOHC). Hydrogen can be stored underground, like natural gas, in empty gas fields, salt caverns, or rock domes. Above-ground, you can store hydrogen in tanks (hydrogen that is compressed, liquefied, bound to a carrier like LOHC, locked in a molecular structure, etc.). Large-scale, long-term storage of hydrogen plays an important part in balancing a sustainable hydrogen and electricity system. Parts of a sustainable hydrogen system are the production of hydrogen through the conversion of an energy source into hydrogen, and (possibly) the extraction of hydrogen. The large-scale conversion to hydrogen without fossil CO_2 emissions can be done in three ways: (1) with sustainable and nuclear electricity via the electrolysis of water, (2) from natural gas via methane pyrolysis, and (3) from biomass by means of various conversion technologies. Large-scale production will mostly take place next to the source, at locations where it can be done cheaply. Small-scale hydrogen production is possible using solar, wind, or energy from biomass residues, and will be done locally for the most part.

Conversion into energy carriers

To be able to use energy from sustainable sources, we must first convert it into a manageable form: an energy carrier. Which conversion technologies do we have for this purpose, and where does the conversion take place?

We want to move toward an energy system that makes use of sustainable energy sources. Virtually all of these sources can be traced back to the sun. The solar energy that reaches the Earth averages about 340 watts per square meter [3]. This energy from the sun is also the origin of wind (through air pressure differences), the evaporation of water, and the growth of plants. The sun is therefore the source of wind and wave energy, energy from hydro power and biomass, energy from temperature differences in the ocean, and energy from differences in salt concentrations between freshwater and saltwater. Only two sustainable energy sources cannot be attributed to solar energy: tidal energy (generated by the gravitational pull of the moon and the sun) and geothermal energy (generated by radioactive decay in the Earth's core). The potential of these two energy sources is very small compared to those energy sources from the sun.

The sun is the source of virtually all sustainable energy, such as solar, wind, wave, hydro, and biomass

From sustainable source to usable energy

Solar, wind, hydro, and biomass are sources of sustainable energy. In order to use sustainable energy, we first need to convert the sources into energy carriers (such as electricity), gas (such as hydrogen and methane), heat, or into feedstocks like hydrogen (H_2), carbon (C), carbon dioxide (CO_2), and oxygen (O_2). Energy carriers allow us to deliver energy services, like heating, cooling, and lighting in homes. With feedstocks we can make materials and products, such as plastics or fertilizers. Numerous conversion technologies have already been developed for the conversion of sustainable energy into useful energy carriers and feedstocks. Some of them are already mature, while others are still in the research stage (R&D).

The table on p. 17 shows that sustainable energy sources allow you to produce various energy carriers: electricity, hydrogen, heat, and cold. And that you can convert biomass residues not only into energy carriers, but also into feedstocks (carbon in various forms). The electrolysis of water not only generates hydrogen—an energy carrier as well as a feedstock—but also a second feedstock: oxygen, the largest portion of which will actually be released into the air.

Sustainable energy sources and conversion technologies

Source	Conversion process/technology	Maturity*	Output
Sunlight	Solar collector	Mature	Heat
Sunlight	Solar cell (PV) Concentrating solar power (CSP)	Mature Mature	e^- (Electricity) e^- + Heat
Sunlight	Solar cell (PV) on surface waters Solar cell (PV) on oceans	Mature First plants	e^- e^-
Sunlight + water	Solar cell with integrated electrolyzer CSP with integrated electrolyzer Solar cell with integrated electrolyzer on oceans	Engineering R&D R&D	$H_2 + O_2$ $H_2 + O_2$ $H_2 + O_2$
Sunlight + water	Photo-electrochemical cell	Prototype	$H_2 + O_2$
Wind	Onshore wind turbine Offshore wind turbine Floating wind turbine Stationary kite	Mature Mature Near mature Pilot	e^- e^- e^- e^-
Wind + water	Onshore wind turbine with integrated electrolyzer Offshore wind turbine with integrated electrolyzer Floating wind turbine with integrated electrolyzer Kite ships	Near mature Engineering Engineering Prototype	$H_2 + O_2$ $H_2 + O_2$ $H_2 + O_2$ $H_2 + O_2$
Hydro power	Turbine	Mature	e^-
Hydro power + water	Turbine with integrated electrolyzer	Mature	$H_2 + O_2$
Ambient heat + Renewable e^-	Heat pump	Mature	Heat
Geothermal	Heat exchanger	Mature	Heat
Geothermal	Steam turbine	Mature	e^- + Heat
Tidal energy	Turbine	Mature	e^-
Seawater air conditioning (SWAC)	Deep cold seawater with heat exchanger for direct cooling	Mature	Cold
Blue energy (osmosis)	Salinity gradient power using membranes	Pilot	e^-
Waves	Wave energy convertor	Pilot	e^-
Ocean currents	Current turbine	Pilot	e^-

Read more on page 18 >

Source	Conversion process/technology	Maturity*	Output
Ocean thermal energy conversion (OTEC)	Thermal gradient power using low temperature cycles (Rankine cycle)	Pilot	e^-
Solid biomass, biogenic waste	Boiler, furnace, stove	Mature	Heat
Solid biomass, biogenic waste	Steam turbine	Mature	e^- + Heat
Solid biomass, biogenic waste	Gasification Plasma gasification	Mature First plants	$H_2 + CO_2 + C$ $H_2 + CO_2$
Solid biomass, biogenic waste	Digestion + boiler	Mature	Heat + CO_2
Wet biomass, biogenic waste	Digestion + gas engine/turbine	Mature	e^- + Heat + CO_2
Wet biomass, biogenic waste	Digestion + steam methane reforming Supercritical water gasification Digestion + methane pyrolysis Microbial electrolysis cell	Engineering First plants R&D R&D	$H_2 + CO_2$ $H_2 + CH_4 + CO_2$ $H_2 + C$ $H_2 + CH_4$

* Development phases: research and development (R&D) → prototype → engineering → pilot → first plants → near mature → mature.

The fact that these conversion technologies also produce feedstocks in addition to energy carriers, creates opportunities for chemical production processes. The most important building blocks for the chemical industry are hydrogen, carbon, oxygen, and nitrogen. Nitrogen is the only one of these feedstocks that cannot be generated through an energy conversion. But one can easily extract nitrogen, as well as oxygen, from the air using an electric air separation unit, and electricity is of course a product of the conversion of sustainable energy.

The fact that the conversion of sustainable energy also produces feedstocks, creates opportunities for making chemical products and synthetic fuels

Fossil versus sustainable energy sources

There are important differences between fossil and sustainable energy sources, for instance, concerning their origin, available quantity, and sites.

We extract fossil energy from the underground, from oil and gas fields, and coal mines; virtually all sustainable energy comes from the sun. We harvest this energy directly (solar energy), from the atmosphere (wind energy), from the hydrosphere (tidal energy from the oceans, hydro power from reservoirs), and from the biosphere (seaweed or wood waste, for example). We extract fossil energy at specific locations, such as the oil fields in Saudi-Arabia, where the ground contains large concentrations. Solar and wind energy can be produced all over the world, nearby or far away from the end-user, but there are significant spatial disparities in the energy density: there is more solar energy in the Sahara than at the north pole.

The sun at the North Pole is never high in the sky

Comparison of some characteristics of fossil and sustainable energy sources

Characteristic	Fossil energy sources	Sustainable energy sources
Origin	Geosphere	Sun, atmosphere, hydrosphere, biosphere, geosphere
Quantity	Finite reserves	Inexhaustible source
Sites	Concentrated in limited number of locations	All over the world, but unevenly spread
Land surface area required	Limited, for oil and natural gas extraction Extensive, for coal extraction	Extensive, for conversion
Time lapse	Continuous output	Fluctuating output

You calculate the efficiency of electrolyzers on the basis of a fuel's HHV (Higher Heating Value) and not its LHV (Lower Heating Value)

Power and use

W	Watt	1 watt (= 1 joule per second)	= 0.001 kW
kW	Kilowatt	1,000 watts	= 1 kW
MW	Megawatt	1,000,000 watts	= 1,000 kW
GW	Gigawatt	1,000,000,000 watts	= 1,000,000 kW
TW	Terawatt	1,000,000,000,000 watts	= 1,000,000,000 kW
Wh	Watt-hour	1 watt-hour (= 3,600 joules)	= 0.001 kWh
kWh	Kilowatt-hour	1,000 watt-hours	= 1 kWh
MWh	Megawatt-hour	1,000,000 watt-hours	= 1,000 kWh
GWh	Gigawatt-hour	1,000,000,000 watt-hours	= 1,000,000 kWh
TWh	Terawatt-hour	1,000,000,000,000 watt-hours	= 1,000,000,000 kWh

The power of an appliance or installation is expressed in watts (W), that is, the energy (in joules) that it can provide per second. Say, for instance, that a microwave oven has a power of 1,000 W, this means that it uses 1,000 joules every second. The greater its power capacity, the faster it will heat up your meal. We express the amount of delivered or used energy in kilowatt-hours (kWh). If a washing machine, with a power capacity of 1 kW (1,000 watts), is operated for one hour, then its energy use is 1 kWh, which is equivalent to 3,600,000 joules.

How efficient is an electrolyzer?

To answer this question, we first need to take a detour and look at heating boilers. You would think that a boiler's efficiency would be expressed as the amount of heat it produces, divided by the energy content of the fuel used. But since, in the case of the old-fashioned boilers, part of the heat disappeared with the flue gases through the chimney, one did not use the total chemical energy content—or the Higher Heating Value (HHV)—of a fuel, but the total minus the condensation heat from the water vapor present in the flue gases. This corrected chemical energy content is known as the Lower Heating Value (LHV). Because modern heating boilers recover the heat from the flue gases, the efficiency of heating boilers now actually exceeds 100%, if you calculate it on the basis of the LHV. This obviously makes no sense. In reality the efficiency of these boilers, when you calculate using the HHV, both for natural gas- and hydrogen-fired boilers, is about 97%.
As the difference between the HHV and LHV increases which means that the HHV/LHV ratio has a higher value, the error in the efficiency calculation based on the LHV increases as well. So, the real energy content of 1 kg of hydrogen is 39.4 kWh, the HHV of hydrogen. But the energy content, expressed in the LHV, is only 33.3 kWh.

Relation between Higher Heating Value (HHV) and Lower Heating Value (LHV) for a number of fuels [1]

Fuel	HHV/LHV
Coal	1.03
Crude oil	1.06
Natural gas	1.10
Hydrogen	1.18

And since hydrogen has the highest HHV/LHV, this distortion is greatest for hydrogen.
But now the efficiency of electrolyzers is also frequently expressed on the basis of hydrogen's LHV. This means that the real efficiency is considerably underestimated, because the calculation does not take into account part of the energy contained in hydrogen. For example: if the LHV efficiency of an electrolyzer is 70%, then the (real) HHV efficiency is 82.8%.

The conversion of solar and wind energy to electricity or hydrogen requires more space than do oil or gas extraction and their conversion into electricity or hydrogen. Thus, a surface area of 80 km^2 is needed for the production of 10 TWh of electricity, using a solar farm set up in the desert, with an operational time of 2,100 hours [4]. While only 1 km^2 is required to achieve the same electricity production with a gas-fired power plant with an efficiency of 60%, and this includes the spatial footprint for the extraction, transport, and the power plants [5]. For its part, the extraction of coal and its conversion into electricity require lots of space, indeed more space than in the case of solar and wind electricity. The production of 10 TWh of electricity with a coal power plant requires about 200 km^2 [6].
A final difference between fossil and sustainable energy relates to the availability in time. Since fossil energy sources are extracted from static (and finite) reserves, their availability practically does not vary in time. But this is certainly the case for solar and wind energy: these two (inexhaustible) sources fluctuate significantly in time, ranging from seconds to seasons.

Hydrogen production with sustainable energy

Now that the conversion of solar and wind power to electricity has become truly inexpensive, its conversion into the hydrogen energy carrier is also interesting. Even if the costs of producing hydrogen through electrolysis will be higher than the cost of producing electricity from the sun, wind, or hydro power, the transport and storage costs of hydrogen are far lower than they are for the same amount of energy in the form of electricity. This means that you can deliver cheap solar and wind electricity, which has been generated far away from the demand, more cheaply, and at the right

TECHNOLOGY FOR THE FUTURE

Reversible fuel cell

A single device with two functions: electrolyzer and fuel cell

Hydrogen is produced by using an electrolyzer to split water into hydrogen and oxygen. The electricity required comes from a sustainable source, for instance, solar or wind. From 9 kg of water, you can make 1 kg of hydrogen and 8 kg of oxygen. In a fuel cell the reverse occurs: 1 kg of hydrogen reacts with 8 kg of oxygen from the air and thereby produces electricity and 9 kg of water. The electrolyzer's technology is the same as that of the fuel cell, the only difference being that the electrochemical process is reversed. You could therefore also use a single device, the operational direction of which you could reverse. Such a device is known as a reversible fuel cell, which you can use for instance to bring electricity supply and demand into balance [8]. But with this technology you can also provide electricity and heat to buildings at all times, even if the supply of solar energy varies significantly. If there is a surplus of solar electricity, the device works as an electrolyzer and produces the hydrogen. In the case of a solar electricity shortfall, at night and in the winter, then the device works as a fuel cell and produces the electricity. In both situations, heat at 60-80 °C is produced, which you can make use of perfectly for heating or to produce hot tapwater [9]. Another context in which such a reversible fuel cell is very useful is in space, for example, when we build a research station on the moon or on Mars [10].

Reversible high-temperature fuel cell that balances electricity supply and demand [11]

time, to the user in the form of hydrogen than in the form of electricity. The additional costs of hydrogen production compared to electricity production is thus outweighed by hydrogen's lower transport and storage costs. We discuss this further in Chapter 2.

Expansion of electrolysis capacity

Electrolysis is not a new process or technology. In fact, electrolysis has been used for about a hundred years to produce chlorine from salt. The salt is first dissolved in water, after which an alkaline electrolyzer is used to split the salt (sodium chloride), thereby releasing chlorine. But because the salt is dissolved in water, the electrolysis also splits the water. Chlorine production thus also generates hydrogen as a "by-product".

Electrolysis has been used for about a hundred years to produce chlorine, generating hydrogen as a by-product

Globally, the electrolysis capacity for chlorine production amounts to about 20 GW. These alkaline electrolyzers are now being modified, and specifically designed, for the production of hydrogen and oxygen from ultra-pure demineralized water.

The production capacity of alkaline electrolyzers is being extended all over the world for the purpose of hydrogen production. Anion electrolyzers are also being developed as an alternative to alkaline electrolyzers, which largely resolves some problems specific to alkaline electrolyzers, such as those concerning the low load and start-up time. Proton exchange membrane (PEM) electrolyzers are also being developed and produced on an increasingly large scale. There are also high-temperature electrolyzers, such as solid oxide electrolysis cell (SOEC) electrolyzers, which, though far more expensive, are suited to specific applications.

Working principle of different types of electrolyzers [7]

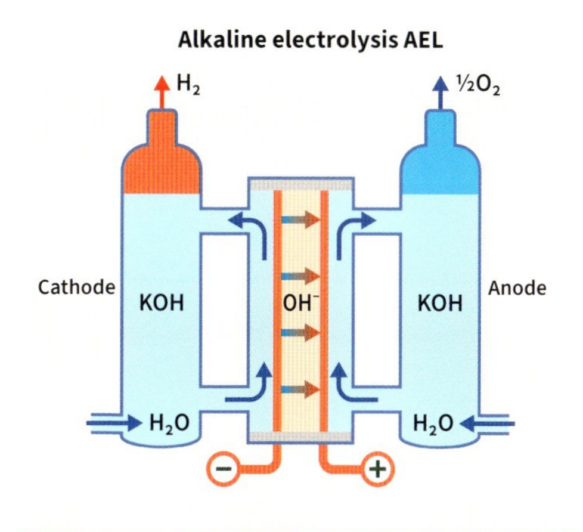

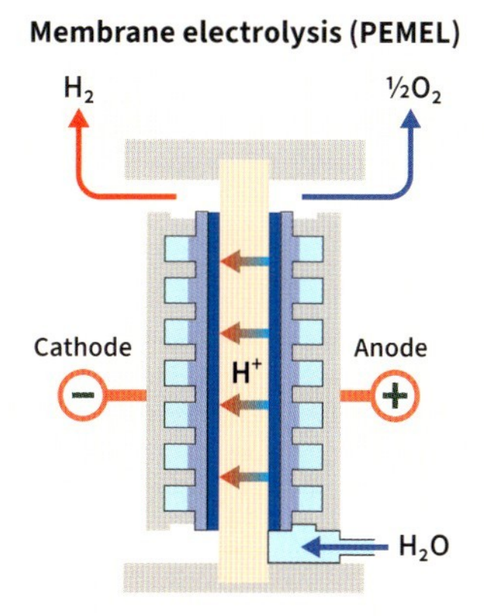

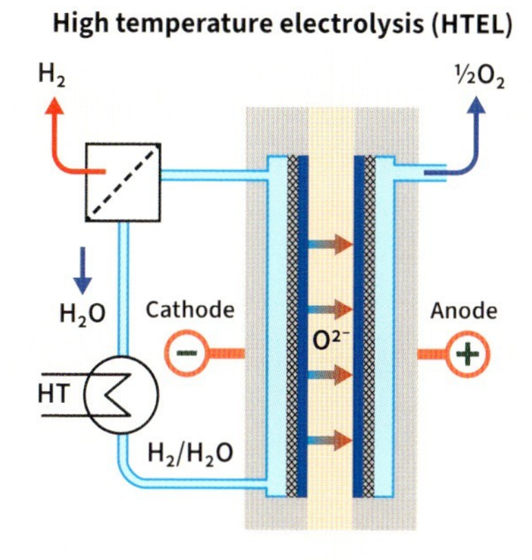

Technology	Alkaline electrolysis (AEL)	Membrane electrolysis (PEMEL)	High temperature electrolysis (HTEL)
Temperature Range	40 - 90 °C	20 - 100 °C	700 - 1,000 °C
Cathodic Reaction (hydrogen evolution reaction)	$2H_2O + 2e^- \rightarrow H_2 + 2OH^-$	$2H^+ + 2e^- \rightarrow H_2$	$H_2O + 2e^- \rightarrow H_2 + O^{2-}$
Charge Carrier	OH^-	H^+	H^+

Decreased prices through electrolyzer cells and stacks mass production

The technology structure of electrolyzers is quite similar to that of solar cells, batteries, and fuel cells. Electrolyzers are comprised of separate electrolytic cells which, for increased output, are piled one upon another in a stack. If more output is desired, more stacks are placed behind each other. The cells have no rotating components, and several commonly used types, including alkaline and PEM electrolyzers, do not make any use of high temperatures. There is therefore no need for costly, heat-resistant materials, and they don't require lots of maintenance. Thanks to this technology structure, electrolyzers are rapidly becoming cheaper through the mass production of cells and stacks, repeating the experience with solar cells and batteries.

Decreased prices through electrolyzer integration

Electrolyzer installations are built with modular units: stacks with a capacity between 1 and 10 MW, which operate on direct current (DC). Solar panels produce DC and, following the first electricity conversion step, wind turbines do so as well. But if you integrate the electrolyzer, by installing electrolysis stacks instead of the DC/AC inverter (direct current to alternating current) in a solar farm or in a wind turbine, then you can avoid several electricity conversion steps. However, the current thinking is focused primarily on setting up large-scale electrolyzer installations seperately at or near large-scale electricity-producing solar and wind farms. The production of hydrogen thus works as follows: in a solar farm, a row of solar panels produces low-voltage DC.

Through mass production of cells and stacks electrolyzers are rapidly becoming cheaper

Transformers in the solar farm then convert this to medium-voltage AC. In the case of a wind farm, the AC produced in every wind turbine is first converted into a low-voltage DC, to then, also in the wind turbine, be converted into medium-voltage AC. The medium-voltage AC is then transmitted using infield cables to a transformer station, where it is converted into high-voltage AC or, in the case of transport over long distances, into high-voltage DC. In a large-scale electrolyzer installation this high-voltage then needs to be again converted into low-voltage DC, because this is what the electrolyzer stacks require. All this involves considerable costs and electricity losses, associated with the various electricity conversions. Indeed, in such a large-scale electrolyzer installation the investment costs for the necessary electricity conversions are even higher than the investment costs for the electrolyzer stacks [12]. As a result

The integration of electrolyzer stacks with solar panels and wind turbines avoids electricity conversion costs and losses

of the integration of electrolyzer stacks with a row of solar panels or a wind turbine, the so-called BOS (balance of system) costs (all costs other than those for the electrolyzer stacks) would be sharply reduced. It is estimated that eliminating the various electricity conversions could result in a decrease in hydrogen production costs of more than € 0.50 per kilo [13].

By integrating the electrolyzer stacks, you in effect develop a solar or wind farm that doesn't produce electricity, but hydrogen. And in doing so, you only need to incur development, engineering, and authorization costs once, rather than twice. Furthermore, this allows you to make standard "products", such as a wind-hydrogen turbine or an electrolyzer stack connected to a row of solar panels. Such product standardization should lower investment costs even more.

White hydrogen

We can use fossil energy or water to produce hydrogen, but hydrogen is also present in the ground at a number of locations. This hydrogen is known as "natural" or "white" hydrogen. Hydrogen can form underground as a result of geo-chemical, chemical, and biological processes, or through radiolysis (the dissociation of molecules through radioactivity). Hydrogen has been found underground at several sites on Earth. In Mali, a small hydrogen source has been used in a gas engine since 2012. But natural hydrogen has also been found underground in other places, including Australia, France, the United States, and Russia. It is still not clear whether large amounts are involved or how we can cleanly and inexpensively extract this hydrogen [14].

Where do you produce which sustainable energy?

Sustainable energy sources are not present everywhere in the world. And they are not equally good everywhere in the world. Nor is there enough available space everywhere to produce sustainable energy. Several forms of sustainable energy can only be produced at specific locations. This limits the application possibilities. Hydro power, for instance, requires not only water, but also a big difference in elevation, as is found in mountainous areas. For blue energy (from osmosis), you can turn to sites where freshwater and saltwater can mix, as in river estuaries in the sea. The extraction of geothermal heat is only possible if the hot water or the heat is not situated too deeply underground, and if there is a concentrated demand for heat in the vicinity. Cooling with seawater can only be done on the coast, in places where the sea quickly becomes both deep and cold, and where there is a concentrated demand for cold nearby.

For the production of solar and wind energy things are a little different. Practically everywhere in the world the sun shines at times and the wind blows on occasion, albeit not consistently. The highest solar irradiation occurs in deserts—usually large, virtually uninhabited areas. The same applies more or less to wind. The highest wind speeds occur on the oceans, where lots of space is available and absolutely nobody lives. You can also often turn to the sun and wind to supply part of the energy requirements closer to where the energy is demanded. But only in the deserts and on the seas and oceans you can really produce sustainable energy cheaply. If you then convert that energy into hydrogen, you can also cheaply transport cheap sustainable energy over long distances and store it in large volumes.

By integrating electrolyzer stacks with solar panels and wind turbines, you develop an integrated solar- or wind-hydrogen farm

Geothermal power plant in Iceland

Sustainable energy production from biomass mainly concerns energy production from biomass residues, both wet (such as manure, sludge, and fruit and vegetable waste) and dry (such as wood waste residuals, sawdust, and straw). These residual streams are found at production sites and processing industries. Thus cows, pigs, and chickens produce manure in the stalls at livestock farms, while sawmills generate sawdust as their residual. These residual streams can be processed on a small-scale on site, but it is often economically more attractive to collect and process them on a somewhat larger scale. Food consumption and the use of biomaterials, like wood, paper, and cardboard, also generate residual streams. These are frequently small and widely dispersed. The exploitation of this type of biomass as an energy source on a large scale therefore calls for an intricate collection system. Besides the use of biomass residues, the cultivation of seaweed for food, energy, feedstocks, and materials production is also being examined. But seaweed farming takes place at sea.

Only in deserts and on oceans you can really produce solar and wind energy cheaply

In summary, you can say that the production of sustainable energy is, for different reasons, closely linked to specific locations and thus varies in time. Many sources present a big and inexpensive potential, but this energy supply is located very far away from the energy demand. Transport and storage are therefore of fundamental importance in a sustainable energy system.

Sustainable energy system goals

Every energy system must ensure that people and companies can purchase the energy services they desire at the right moment. But what should an energy system actually achieve? And what is the difference between a fossil and a sustainable energy system?

The assessment and comparison of energy systems are often based on their goals. In the case of a fossil energy system, it was primarily a matter of being “affordable”, “reliable”, “energy-supply secure” and “safe”. But additional goals have been added for a sustainable energy system, namely, “clean”, “circular”, “materials-supply secure” and “fair”. Many of these goals cannot be achieved completely. You can never realize an energy system that is 100% reliable or 100% safe. You can however formulate these goals in quantitative terms.

Goals of a sustainable energy system

Clean

An energy system is clean if no or only low emissions of pollutants are released into the air, water, or soil. Many emissions are subject to legal standards, expressed in terms of emission amount per unit of energy production or unit of energy use. An energy system can also have as a goal the reduction of the total amount of emissions to a specific level over a specific period of time.

The emission reduction of greenhouse gases responsible for climate change is currently an important global environmental goal. The most important greenhouse gas is CO_2, which is released by the combustion of fossil fuels. Other gases that contribute to strengthening the greenhouse effect are—in order of increasing importance—methane (CH_4), laughing gas (N_2O), CFCs, HCFCs, and HFCs [15]. Specific greenhouse gas emission targets have been set, for example, per kilometer driven or per tonne of steel produced. But total greenhouse gas emission targets have also been defined, such as "total greenhouse gas emissions must be at least 55% below 1990 levels by 2030".

To achieve a clean energy system we must transition to renewable energy sources and reusable materials

Hydrogen is not a direct greenhouse gas, because the hydrogen molecule has no dipole moment (measure for the polarity of a bond or molecule) that prevents the absorption of infrared radiation. Hydrogen however can work as an indirect greenhouse gas. All indirect greenhouse gas effects of hydrogen are caused by its oxidation in the atmosphere. This oxidation influences the life span of other greenhouse gases in the atmosphere. Hydrogen reacts, among others, with OH-radicals, with the result that these cannot react with methane, which is a powerful greenhouse gas. Even though the combustion or chemical reaction of hydrogen emits no hydrogen, hydrogen leakage can occur at several points in the overall chain. The amount of leakage is usually very small, but attention naturally needs to be paid to its prevention or minimization. On the other hand, hydrogen will replace natural gas (methane), so that less methane will be released into the atmosphere. In net terms, the greenhouse gas effect is therefore smaller [16].

The energy system also releases other substances. Emissions into the air, which occur during the conversion into energy carriers, can in particular cause significant air pollution locally. The combustion of fossil and biofuels, besides CO_2, also releases nitrous oxides (NO_x) and particulate matter. In principle, the consumption of electricity and hydrogen produces no local emissions of CO_2, NO_x, and particular matter. But this only applies if, in the conversion of hydrogen into electricity and heat, you use a fuel cell—an electrochemical conversion, whereby chemical energy is directly converted into electricity and heat. If, instead, you combust the hydrogen, NO_x will be released, but no CO_2 or particulate matter. Electricity and hydrogen are also known as "carbon-free" energy carriers, because no CO_2 is released when they are used.

In addition to air pollution, emissions can occur into surface water and the soil. Thermal electricity production for example involves the use of large volumes of cooling water, which is usually then discharged into surface water. And the extraction of fossil fuels and raw materials requires lots of water, and frequently chemicals as well. After use, the polluted residual streams are discharged or dumped.

Such emissions can often be cut back with so-called "end-of-pipe" technologies. For instance, catalytic converters reduce automobile NO_x emissions, filters reduce particulate matter emissions, and water treatment plants effectively treat wastewater so that it can be safely discharged into surface water. But preference should be given to preventing emissions at the source and during the conversion. End-of-pipe solutions always involve a great deal of uncertainty as to the actual emission reduction.

Solar boilers are used in the Himalayas to cook food and boil water

In practice, the technologies used are not always effective enough, and in many instances not even applied at all. To achieve a clean energy system, we must therefore ultimately transition to renewable energy sources and reusable materials. Only then fewer emissions will occur at the source. However, since emissions occur even in a sustainable energy system, attention will always need to be directed at emission reduction, in all components of the energy system.

Affordable

An affordable energy system is a precondition for economic development and a good quality of life. The rapid economic development over the last century was in part made possible by the availability of cheap fossil energy: crude oil, natural gas, and coal. Conversion technologies and transport networks ensured that this energy was available in the form of usable energy carriers (fuels, electricity, and gas) to all economic sectors: from agriculture to mineral extraction, from chemical companies to food producers, and from restaurant owners to transporters. Affordable energy also improves the population's quality of life, because it allows us to heat, cool, and light our homes, and to transport ourselves and all kinds of goods easily and cheaply.

The negative side of these developments is that they have been associated with a great increase in the pollution of our living environment, and with climate change. And this now threatens to degrade the quality of life. By switching over from a fossil to a sustainable energy system, it will be possible to preserve the quality of life in a manner that

Cost development for solar PV, concentrating solar power (CSP), onshore wind, and offshore wind [17]

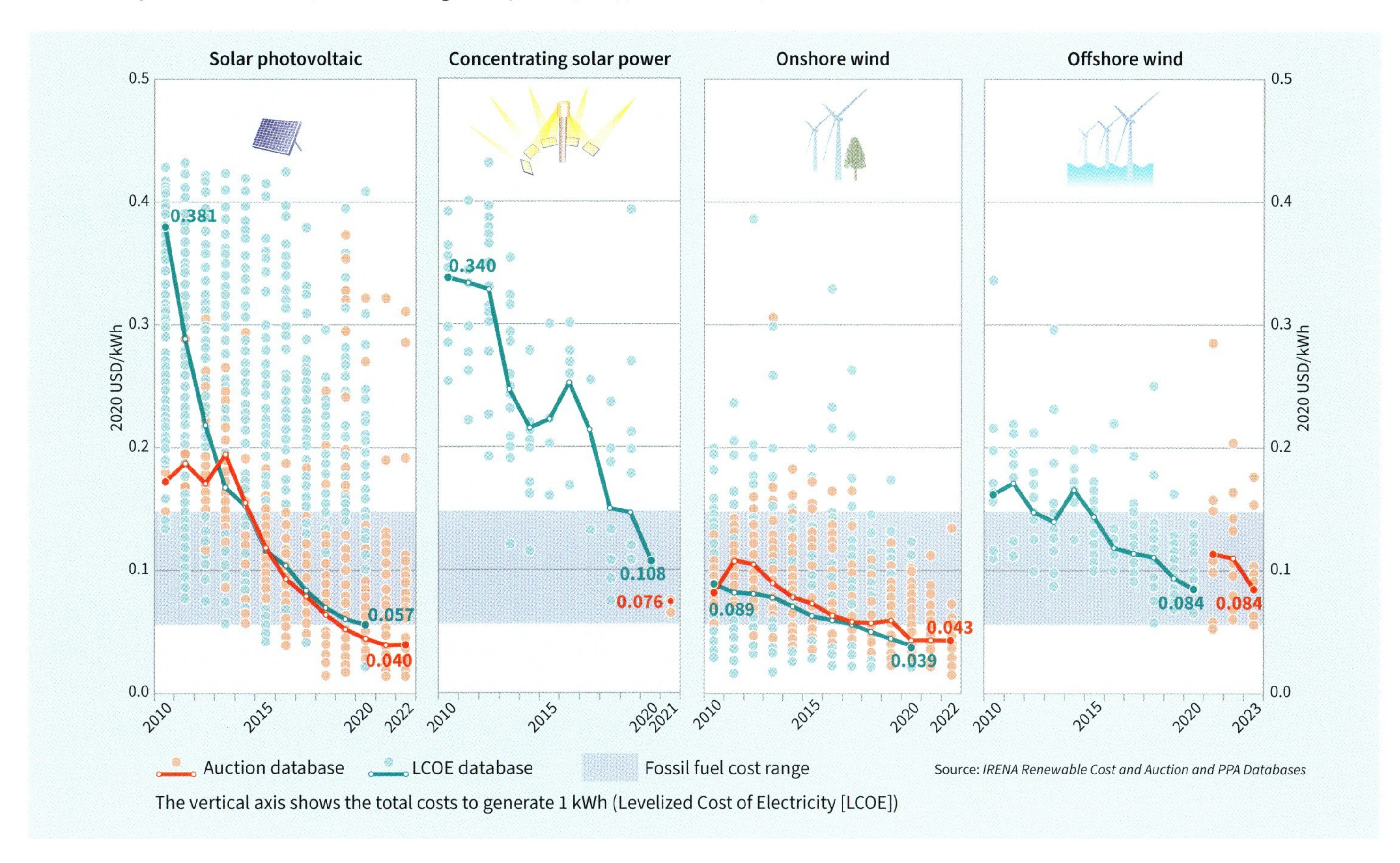

no, or at least a lot less, damage is inflicted on our living environment. The question is whether this is possible without incurring additional costs. The calculations above suggest that this is possible in principle.

You can, however, only produce cheap electricity from the sun and wind at locations where there are, respectively, lots of solar irradiation and high wind speeds, as well as a great deal of inexpensive space. These places are often situated at a great distance from the locations where energy demand is high. In order to cost-effectively store and transport the sustainable electricity over great distances, you first convert it into hydrogen. This represents an extra cost item. But if a large-scale hydrogen market were to develop, you could produce hydrogen at the good locations for € 1.00-1.50/kg. This hydrogen price is at the same level of a price of natural gas of € 0.25-0.38/m^3, or € 0.03-0.04/kWh [18], [19], [20], [21], [22].
The storage and transport of hydrogen is many times cheaper than the storage and transport of electricity. You can transport hydrogen for € 0.40-0.50/kg over a few thousand kilometers via pipelines. Added to the production price of € 1.00-1.50/kg, the total cost of producing, storing, and transporting hydrogen amounts to € 1.40-2.00/kg. This corresponds to a € 0.35-0.50/m^3 price of natural gas, or € 0.035-0.05/kWh. This hydrogen is not only affordable, but also available day and night, summers and winters. And its price is competitive with that of fossil energy plus a CO_2 tax.

In many countries imported hydrogen will start competing with locally produced hydrogen, and even solar and wind electricity, for example, in the mobility and industry sectors (including for the production of high-temperature heat). In addition, imported green hydrogen (produced without CO_2 emissions), which can be stored on a large scale and cheaply, will be used to produce electricity when sustainable electricity cannot be produced in sufficient amounts locally. Such a scenario is also conceivable for space heating: electric heat pumps provide the baseload heat, while hydrogen can take care of the winter heat-demand peaks.

From the above you can conclude that a sustainable energy system will become affordable in the future. The production costs of sustainable energy will possibly drop over time, but the transport and storage costs are higher than those for fossil energy. However, by using hydrogen instead of electricity, you can sharply reduce the transport and storage costs of sustainable energy.

Reliable

An energy system is reliable when it can supply the demanded amount of energy at all times to end-users, and with the right quality. This reliability concerns: (1) the supply of the demanded amount of energy of the right quality from production or storage, and (2) the transport and distribution of the energy to end-users. Both elements are subject to failure, so that a 100% reliable energy system is in principle not conceivable.

Energy demand varies not only over the day and with the seasons, but also with work and production times

The demand for energy varies. Thus, the energy demand for space heating in Japan and in Northern and Central Europe in the winter is greater than in the summer. In southern countries energy demand for cooling will be greater than for heating, and is higher in the summer than in the winter. The demand for energy for lighting also has a clear seasonal pattern: it is lowest in the summer, and highest in the winter. Energy demand varies not only with the seasons, but also with work and production times.
In the basic metal and basic chemicals sectors, in which production is continuous, so too is the demand for energy. Such continuous demand is known as the baseload.
In offices, energy demand during working hours (during the day, on working days) is greater than at other times.
A reliable energy system has sufficient capacity and is adjustable, so that the production, import, and storage of energy can be aligned to these fluctuations in demand.

A reliable, sustainable energy provision requires large-scale storage of hydrogen or hydrogen products

In a fossil energy system, the storage of energy is simple and cheap: coal in the open air, oil in tanks, and gas in empty gas fields and salt caverns. This makes for high production and delivery reliability. A fossil electricity system, to be sure, must have a certain extra capacity, since power plants can fail or stop production temporarily for maintenance work.

In a sustainable energy system the storage of energy is less simple, certainly if the production of the sustainable energy mainly uses solar and wind. You can't store electricity that easily, and certainly not on a large scale in order to accommodate seasonal fluctuations. In such a system you not only need to deal with fluctuations in energy demand, but also with fluctuations in energy production. If there is not enough sun and wind to produce sufficient electricity, you can complement the production with hydro power, geothermal energy, and hydrogen.
A reliable sustainable energy provision therefore requires the large-scale storage of hydrogen or hydrogen products. We can thus store hydrogen in the same manner as natural gas. This hydrogen storage is necessary to enable the reliable supply of electricity as well as hydrogen to users.

Materials used in cars and electricity production [24] *

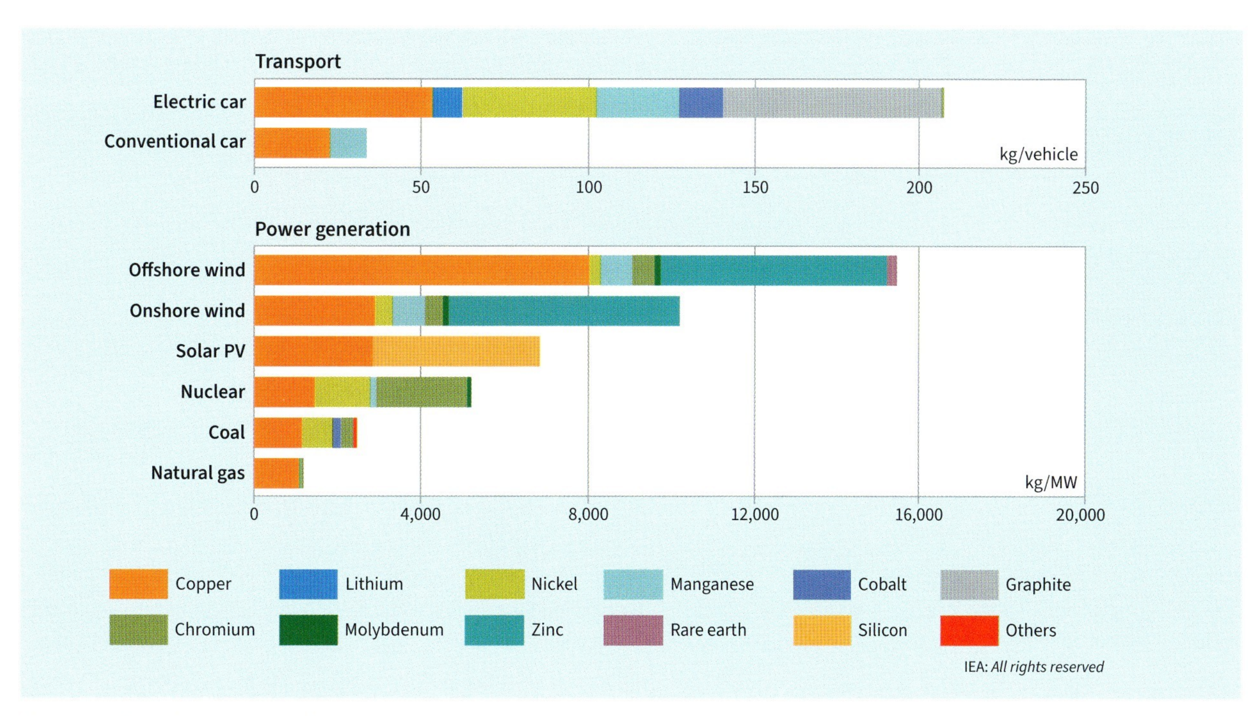

* Steel and aluminium are not included.

Energy is delivered by means of an infrastructure. This can be an electricity or gas infrastructure, but also be an energy transport infrastructure by road or over water. These infrastructures contain lots of redundancy—their dimensions are larger than strictly necessary, so that delivery can continue even if a section fails. A good example of this are the loop systems, through which energy can be supplied via different routes. This ensures the reliability of the energy delivery.
Ultimately, we want the reliability of the infrastructure and of the delivery in a sustainable energy system to be comparable to that of a fossil energy system.

Circular

An energy system is circular when the materials used are extracted and produced in a clean manner, when these materials are used efficiently, and when the system reuses these materials. “Materials” here refers to raw materials that are extracted from the ground, including metal ores (like copper and zinc), mineral rocks (like silicon and graphite), and a large group of rare-earth elements.
The reserves of these materials are finite, as in the case of fossil fuels. A sustainable energy system uses many more materials than does a fossil energy system. The energy transition is therefore not only a transition from fossil energy to renewable energy, but also a transition from fossil energy (crude oil, natural gas, and coal) to materials (metals, minerals, and rare earths) [23]. This last group is referred to in short as “materials”.

To produce solar cells, wind turbines, and batteries, you need more materials than you do for power plants and storage tanks. For an electric car you need six times more minerals compared to a gasoline- or diesel-powered car, and for an onshore wind farm nine times more than for a gas-fired power plant.

Each technology makes use of different materials. Lithium, nickel, cobalt, manganese, and graphite are essential for batteries, although work is being done on batteries without cobalt; rare-earth elements are essential for the permanent magnets in wind turbines, and in the electric motors of battery electric vehicles [25]. Today's solar panels contain silicon, aluminium, and silver [23]; electric grids contain enormous quantities of copper and aluminium. For their part, transformers use lots of copper [25], while alkaline electrolyzer electrodes are made primarily of nickel and aluminium, and PEM electrolyzer electrodes of carbon, iridium, and platinum [26].

The energy transition is also a transition from fossil energy to materials

The transition to a sustainable energy system will lead to an enormous increase in the use of materials. In meeting the objectives of the Paris Agreement, clean energy technologies will account for a sharply growing share of the total materials used over the coming decades: to more than 40% of the total for copper and rare-earth elements, to 60-70% for nickel and cobalt, and to almost 90% for lithium. Battery-equipped electric vehicles and battery storage are already the biggest users of lithium (a position held until recently by consumer electronics), and they are also rapidly on their way to becoming the biggest users of nickel (a position currently still held by stainless steel [24]).

To limit as much as possible the general use of materials, and of rare-earth metals in particular, the following principles could be applied:

- **Refuse, Reduce,** i.e., decrease use, increase efficiency
- **Replace** critical materials with less critical materials
- **Repair, Refurbish, Reuse** components or parts that incorporate materials
- **Recycle** materials at the end of their life span

These principles are in themselves all necessary, but recycling frequently means that the quality of the recycled material is lower. To enable recycling while preserving quality, other product designs and different kinds of materials use are needed. It is also important that the extraction and/or production of materials be carried out in a clean manner, without emitting greenhouse gasses and polluting substances, and without wasting water.

Energy-supply secure

An energy system must ensure that there is sufficient security in terms of access to and supply of energy. Particularly in developing countries, access to energy is now a major problem. Security of energy supply concerns two key questions: (1) Is the world's total supply of energy sufficient? and (2) Is the energy supply evenly and fairly distributed around the world? Natural disasters and geo-political conflicts can disrupt the security of energy supply in specific regions and countries.

Electric vehicles and battery storage are already the biggest users of lithium, and will shortly also be of nickel

Access to energy, and especially to clean, safe, and affordable energy, is a major problem for many people in developing countries. In 2019 more than 2.6 billion people in the world had no access to clean and safe fuel for cooking. Over 750 million people had no access to electricity, and were not connected to an electric grid [27]. In addition, for many more people the provision of electricity and fuels is very uncertain. The United Nations is attempting to change this through large-scale programs. Modern technologies for sustainable energy conversion, storage, and transport can make a contribution in this effort, so that everyone gains access to clean, safe, and affordable energy.

The amount of fossil fuel energy on Earth is finite. One might therefore think that the amount of energy reserves would present a problem, particularly regarding fossil energy sources. But the depletion of these reserves is not yet in sight, and for the time being fossil energy sources can provide for our energy needs. In other words, we are switching over to sustainable energy sources not because the fossil energy sources are depleted, but because the fossil energy system is too great a climate and environmental burden.

Sustainable energy sources are infinite, but not directly usable. Before you can use them, they need to be converted into a usable energy carrier, like electricity or hydrogen. To accomplish this on a global scale, the production and installation of solar cells, wind turbines, batteries, and electrolyzers have to be expanded on a gigantic scale. Moreover, there is more than sufficient space available worldwide for the production of energy from the sun and wind.

Areas with high wind speeds or high solar irradiation are not equally spread over the world. Nor does every place in the world have enough space available for sustainable energy production. Deserts and oceans are eminently suited for the inexpensive production of lots of sustainable energy: they have plenty of space and sun and wind. But because the demand for energy in these areas is limited, any energy produced in them needs to be stored and transported to areas that do have a strong energy demand.

Geopolitics can have a strong impact on the import and export of energy. The oil producing countries for instance, united in OPEC, jointly determine how much oil is produced, giving them a strong influence on the price of oil. In the 1970s OPEC declared an oil embargo on Western countries on one occasion, and the Iran-Iraq war sharply impacted oil supplies. In response, the Western countries, united in the International Energy Agency (IEA), built up strategic oil reserves in all countries equivalent to 90 days net annual oil imports. These oil stockpiles can be drawn on in the event of major oil supply disruptions.

In the deserts and on the oceans there is sufficient space, and sun and wind, to cheaply produce lots of electricity and hydrogen

Energy supply can also be disrupted in a sustainable energy system. Not only as a result of geopolitical acts, but also because of natural disasters and wars. To deal with such eventualities, it is imperative that a strategic hydrogen reserve be established, as has been done for oil. The size of the stockpile could be just as large, namely, the equivalent of 90 days net annual hydrogen imports.

Materials-supply secure

For a sustainable energy system, the security of materials supply is perhaps even more important than that of the energy supply. Sun and wind are actually infinite sources, while the amount of materials on Earth is finite. Indeed, some materials are already scarce today. The extraction and production of many materials is concentrated in a limited number of countries; these are actually fewer in number than the oil and natural gas producing countries. More than three-quarters of all lithium, cobalt, and rare-earth elements is produced by only three countries. In the case of some materials, a single country is responsible for about half of the world's production. Thus, in 2019, the Democratic Republic of the Congo (DRC) and China mined, respectively, 70% and 60% of all the ores containing cobalt and rare-earth elements. The situation is even more extreme in the case of the extraction of materials from the ores: China alone produces almost 90% of all rare-earth metals [28].

In view of the small number of producing countries, the security of the materials supply could become significantly affected by geopolitics. In order to safeguard the security of materials supply, it makes sense for countries that develop sustainable energy technologies to consider taking one or more of the following measures:

- If possible, extract and produce mineral resources themselves.
- Establish a mineral resources bank, which holds the minerals and "lends" them out to sustainable energy technology producers.
- Build up strategic reserves of critical mineral resources, comparable to the current strategic oil reserves.

Safe

An energy system is considered safe when as few people as possible suffer accidents, injuries, or health damage during the generation, transport, storage, and use of energy. For the fossil energy system, there are laws, regulations, standards, and authorizations to ensure this safety. For the sustainable energy system this is not yet, or not entirely, the case.

Of course, new technologies must also be safe, and their safety should ultimately be ensured in the same manner that it is today for existing technologies. But if we start off by setting out all the regulations, laws, and standards, and only then get down to implementing sustainability, then it will take us far too long. In the initial period, it would for instance be desirable to leave room for experimentation, or a more process-oriented approach ought to be pursued, for which a committee of experts could for example grant an authorization or approval, and which would be monitored, and the standards would then ultimately be established on the basis of practical experience.

Fair

An energy system is fair when the benefits and burdens of energy production and use are fairly distributed among different groups of people and countries. In the assessment of a system's fairness, the following themes play a role: energy access, energy poverty, energy taxation and subsidy distribution, influence on energy-policy decision-making, and influence on investment decisions. The fairness can be assessed at a national, regional, and global level; here, we address the fairness at the global level.

Today's fossil energy system involves large import and export flows of crude oil, coal, and natural gas. The industrialized countries import lots of fossil energy, while a limited number of countries export large amounts of fossil energy. The returns from the exploration and export of fossil energy are not very fairly distributed globally: only a small number of energy companies and countries earn a great deal from these activities.

The initial situation for the development of a sustainable energy system differs from one region to another. Europe, Japan, South Korea, parts of the Unites States, and parts of China have insufficient space and insufficient sources of sustainable energy to produce enough affordable, sustainable energy within their own borders. Other parts of the world, such as The Middle East, Australia, New Zealand, large parts of South America and the deserts in Africa possess so many good sources of sustainable energy and so much space, that they could even become exporters of sustainable energy—in the form of hydrogen or hydrogen-derived products for example. There are also countries, like India and Indonesia, that actually have good sources of sustainable energy, but do not have enough space to produce sufficient sustainable energy themselves. The reason is their high population density. It is anticipated that many industrialized countries will start importing mainly sustainable energy, while many developing countries could begin exporting mainly sustainable energy. A fair energy system for the world will equitably distribute the benefits and burdens between these future sustainable energy exporters and importers.

Cobalt mine in Congo

Let's take a look at the example of a sustainable energy system for Europe and Africa. Europe cannot itself produce enough sustainable energy. It has too few good sources and too little space. Africa does have enough of these two elements, and could actually produce so much sustainable energy that enough would be left over for export to Europe and other parts of the world. But Africa lacks the required technology and capital, and a well-trained workforce to independently develop a sustainable energy system. Collaboration in such a situation is advantageous to both continents: the joint development of a sustainable energy system for own-use in Africa as well as for export, would provide Africa with future-oriented economic development, jobs, and welfare. And Europe could realize a clean, reliable, and affordable energy system, on the one hand, by producing sustainable electricity and hydrogen in Europe and, on the other, importing hydrogen from Africa through repurposed natural gas pipelines and by ship. Additionally, Europe and Africa could together build a sustainable circular industry and food production system, both based on solar and wind electricity and hydrogen. Such a collaboration would be fair, from both European and African perspectives [28].

More than three-quarters of all lithium, cobalt, and rare-earth metals is produced by only three countries

Energy efficiency: not a goal, but sometimes a means

Energy efficiency is not an energy system goal in itself, but can sometimes provide a means of making it possible to achieve energy system goals such as being clean, affordable, reliable, or circular. Below, we use two examples to show why increasing energy efficiency does not by definition lead to cheaper energy.

Solar cells, based on silicon, which are installed on homes and in solar farms, convert sunlight into electricity with an efficiency of about 20%. As an alternative, the market offers GaAs solar cells (thin-film cells with gallium arsenide) with an efficiency of about 30%. But we do not install these more efficient cells on our roof. The reason is simple: they are many times more expensive than the solar cells now used. If you were to install these more efficient, but also much more expensive, solar cells on your house, the cost per kWh would rise sharply. GaAs solar cells are applied primarily in aerospace, because of the surface area constraints in that context. In aerospace, the output per surface unit—and thus the efficiency—is clearly primordial.

If you install more-efficient GaAs solar cells instead of silicon ones on your house, the electricity production costs per kWh will be higher

If you install solar cells in the Sahara, they will generate two to three times as much electricity as the same solar cells would in Northern Europe. Transport of solar electricity from the Sahara to Northern Europe can be done incurring a transport and conversion loss of 10-20%. This means that one would still end up with almost twice as much electricity than would be the case if the same solar cells were used to produce electricity here. Nonetheless, we do not choose this "more efficient" option, which requires only half the number of solar panels and therefore requires under half as many materials. The transport of electricity is actually so costly that solar electricity produced in Northern Europe is ultimately less expensive. And even though the sun shines in the Sahara at other times than in Northern Europe, the supply of electricity is still not in balance with the demand.

However, if you convert solar electricity from the Sahara into hydrogen, you can transport and store this sustainable energy far more cheaply, despite the conversion costs and losses. If you transport this hydrogen to Northern Europe, you can then provide electricity whenever it is desired. And with this route the "efficiency" is actually even higher than for solar electricity generated in Northern Europe. Because if solar cells in the Sahara generate two and a half times as much than those in Northern Europe, then—after the conversion into hydrogen, and taking into account the transport losses and the reconversion of hydrogen to electricity—there is still one time left over. Solar electricity from the Sahara versus solar electricity from Northern Europe ends up in a draw when it comes to "efficiency": ultimately both situations require the same number of solar panels. And since the solar power from the Sahara becomes available in the form of hydrogen, you can supply solar electricity to Northern Europe, day and night, summer and winter. The reliability of solar electricity from the Sahara, through the hydrogen route, is therefore much better. And in practice this seems also to apply to the affordability [29]. Ultimately it is thus less a matter of an energy system's efficiency, but of achieving the system's sustainability goals: clean, affordable, reliable, energy-supply secure, circular, materials-supply secure, safe, and fair.

The above examples which show that energy efficiency does not by definition lead to a cleaner, cheaper, and more reliable energy provision, are relevant from the perspective of energy conversion. Naturally, the economical and efficient use of energy is and remains extremely important in the realization of a sustainable energy system. But when we speak of energy efficiency, we mostly mean the energy-efficient (end-)use of energy. This relates to avoiding unnecessary energy use, such as leaving on the lights when you're not in the room, using plastic wrapping when it's not necessary, or leaving devices that you don't use on stand-by. In addition, energy-efficiency improvement in (end-)use is also important, for instance, through the better insulation of buildings and homes, or replacement of incandescent lamps by LED lighting.

Comparison of a fossil and a sustainable energy system

Energy system goals	Fossil energy system	Sustainable energy system
Clean	• End-of-pipe-technology needed to reduce greenhouse gas emissions and other pollution	• Sustainable sources inherently clean (no greenhouse gas emissions) • End-of-pipe-technology needed to reduce other pollution
Affordable	• Fossil energy is cheap • CO_2 pricing raises price • Global transport is cheap • Storage is cheap	• Sustainable energy at good locations very cheap (cheaper than fossil) • More and higher transport costs • More and higher storage costs
Reliable	• Continuous supply of energy • Storage needed due to energy use fluctuations	• Fluctuating supply of energy • Storage needed due to energy use and supply fluctuations
Circular	• Limited use of materials	• Extensive use of finite reserves. Circular use of materials therefore needed to realize sustainable energy system
Energy-supply secure	• Fossil energy reserves in limited number of countries	• Great sustainable energy potential in many countries • Cheap supply in countries with good sustainable energy sources and sufficient space
Materials-supply secure	• Materials use is limited and not critical	• Extensive use of materials, including critical materials • Limited number of countries with supply of some critical materials
Safe	• Safety ensured by laws, regulations, standards, and authorizations	• Arrange safety more quickly, to avoid delay in energy transition
Fair	• Revenues mostly for large companies and limited number of exporting countries	• Joint development by industrialized and developing countries can spread welfare fairly

Difference between a fossil and a sustainable system

In this chapter we have seen that a sustainable energy system has different characteristics from a fossil energy system. This makes it difficult to compare them with each other. Comparisons of the two systems currently still often take the characteristics of the existing, fossil energy system as the starting point. This for example means you look at the energy production volumes per year, the costs of energy production, the annual energy use volumes by sector, and the associated annual CO_2 emissions. This provides a limited picture and can lead to sub-optimal policies for the realization of a sustainable energy system. A comparison that does justice to the characteristics of each energy system, be it fossil or sustainable, could assess whether, and to what extent, an energy system contributes to the goals addressed in this chapter: clean, affordable, reliable, circular, energy-supply secure, circular, materials-supply secure, safe, and fair.

2

THE ROLES OF SPACE AND TIME

We extract coal, crude oil, and natural gas from the ground. Their availability is not susceptible to variations in time and space: the season, the weather, whether it's day or night, the position of the sun and the moon. But this is the case for practically every source of sustainable energy. And this has a big impact on the way you need to establish a sustainable energy system.

The current energy system

Before turning to the future, sustainable energy system, we will first outline here how the current fossil energy system functions: how much energy is extracted, how much energy is used and where, and which worldwide energy flows can connect supply and demand. An overview in numbers and maps.

From energy source to energy service

In the fossil energy system, a production system is dedicated to extracting large volumes of coal, crude oil, and natural gas from the ground. This is done at a limited number of locations, after which these resources are transported by ship or pipeline around the world. The oil is processed in refineries into gasoline, aviation fuel, and diesel. In the vicinity of the demand location, the coal, gas, and oil are converted into electricity. Finally, transport fuels and electricity are distributed, primarily at the regional level.
Electricity, gasoline, aviation fuel, diesel, and natural gas are all energy carriers with a standard quality. Following distribution, they are used in the provision of a range of energy services, such as lighting, communication, a comfortable home or workspace, transport from A to B, product manufacturing, and powering a pump. In addition, the fossil energy sources, that is, hydrocarbon molecules, are used as feedstocks for chemical products, like plastic or fertilizer.

Primary energy use

Fossil energy is actually chemical energy which is stored in molecules: chains of carbon and hydrogen atoms, in other words, “hydrocarbons”. When we refer to “primary energy” in the case of fossil energy and biomass, we mean the energy we extract in raw form, before its conversion into an energy carrier has taken place. While in the case of sustainable energy, such as solar, wind, or hydro, references in energy statistics to primary energy actually relate to the energy (usually electricity) produced after conversion. This in fact renders comparisons between fossil and sustainable energy unfair: it always looks as if the contribution of fossil energy is bigger than that of sustainable energy. Total global primary energy use in 2019 was 606 EJ (1 exajoule is 1018 joules), or 168,000 TWh (1 TWh is 1 billion kWh), of which more than 80% comes from fossil sources [30].

Final energy use

The conversion of primary fossil energy or biomass produces usable energy carriers, such as transport fuels and electricity, and feedstocks like naphtha. The energy

In 2019, 64% of all electricity was generated with fossil energy, 10% with nuclear energy, and 26% with sustainable energy

Global primary energy use in 2019 [30]

Energy source	Share (%)
Crude oil	30.9
Coal	26.8
Natural gas	23.2
Biomass and waste	9.4
Nuclear energy	5.0
Hydro power	2.5
Other	2.2
Total	**100% = 168,000 TWh (1 TWh = 1 billion kWh)**

Global electricity production in 2019 [30]

Energy source	Share (%)
Coal	36.7
Natural gas	23.6
Hydro power	15.7
Sustainable sources (excl. hydro power and waste)	10.8
Nuclear energy	10.4
Crude oil	2.8
Total	**100% = 27,000 TWh**

that is delivered to end-users following the conversion we call the final energy use. Since losses occur in the conversion process (for example, in power plants and refineries), the final energy use is smaller than the primary energy use. In 2019, the total global final energy use amounted to 418 EJ, or 116,000 TWh. This represents 69% of the primary energy use.

Electricity production

In 2019 global electricity production was just under 27,000 TWh, and accounted for about 23% of the final energy use. About 64% of all electricity was generated with fossil energy, 10% with nuclear energy, and 26% with sustainable energy sources.

The supply and demand of electricity need to be in balance at all times. In a fossil energy system, this is done using easily adjustable, fossil-fired power plants. These provide for flexibility in the electricity system. Indeed, the fluctuating supply of solar and wind energy in 2019 was entirely absorbed by fossil-fired power plants and adjustable hydro-power plants.

Primary energy use per inhabitant [31]

Tropic of Cancer

Equator

Tropic of Capricorn

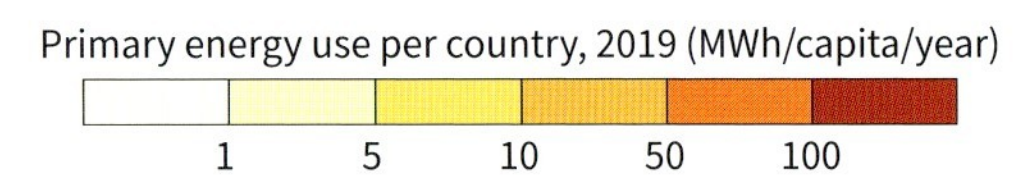

Energy use per capita

Energy use can be broken down in other ways, for instance, on a per capita or per Gross National Product (GNP) basis. These measures often provide a good picture of a country's economic development and prosperity.

Crude oil energy flows around the world [32]

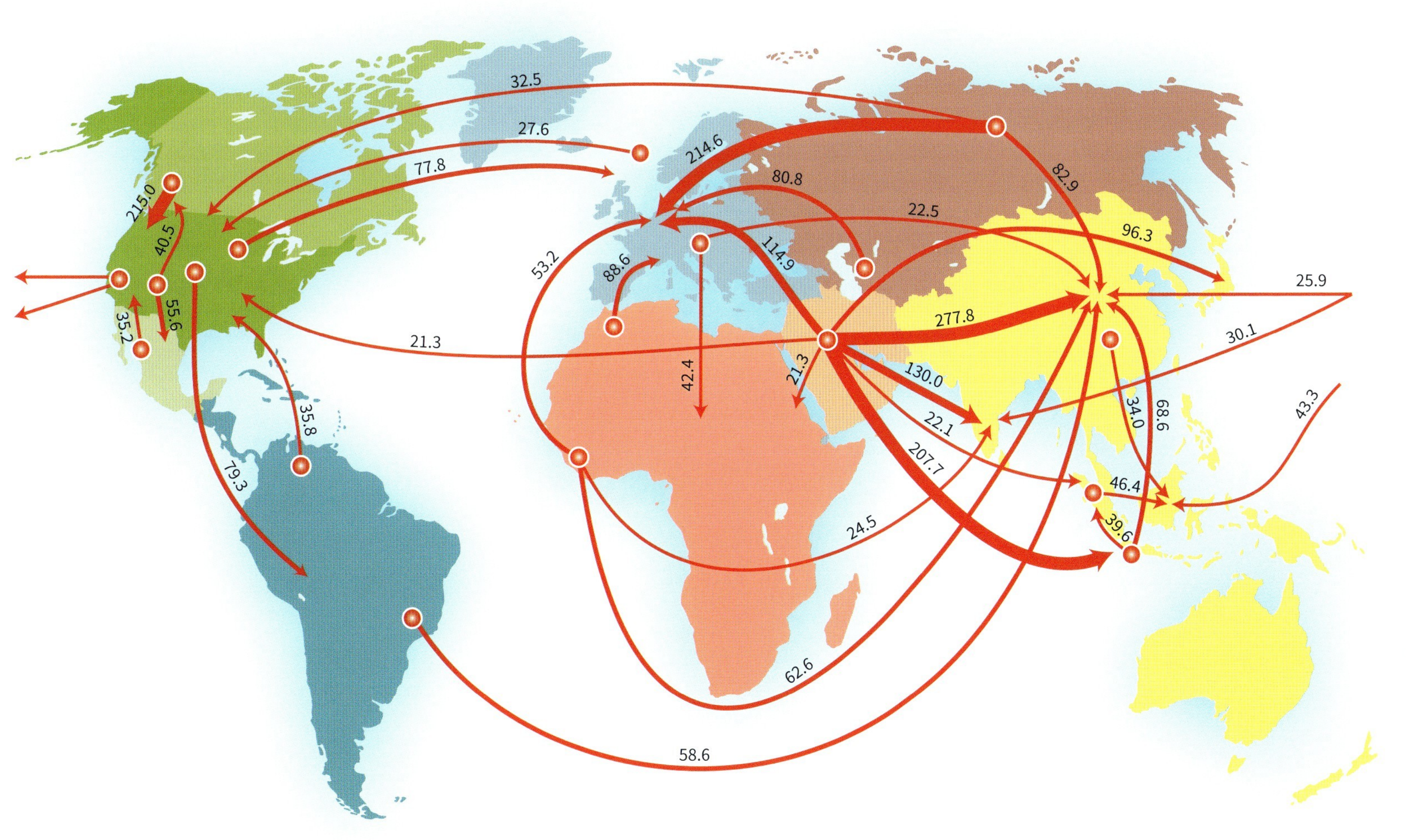

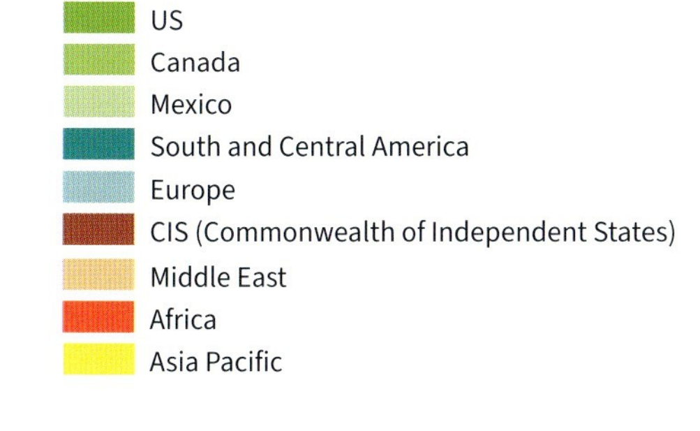

Energy flows around the world

More than 80% of total primary energy use in 2019 originated from fossil sources. Following the extraction of coal, crude oil, and natural gas in a limited number of countries, these energy sources are transported by ship or via pipelines to areas where there is an energy demand. The Middle East and Russia are big suppliers of oil and gas, while Europe, Japan, and Asia are big customers. The United States is essentially self-sufficient.

Natural gas energy flows around the world [32]

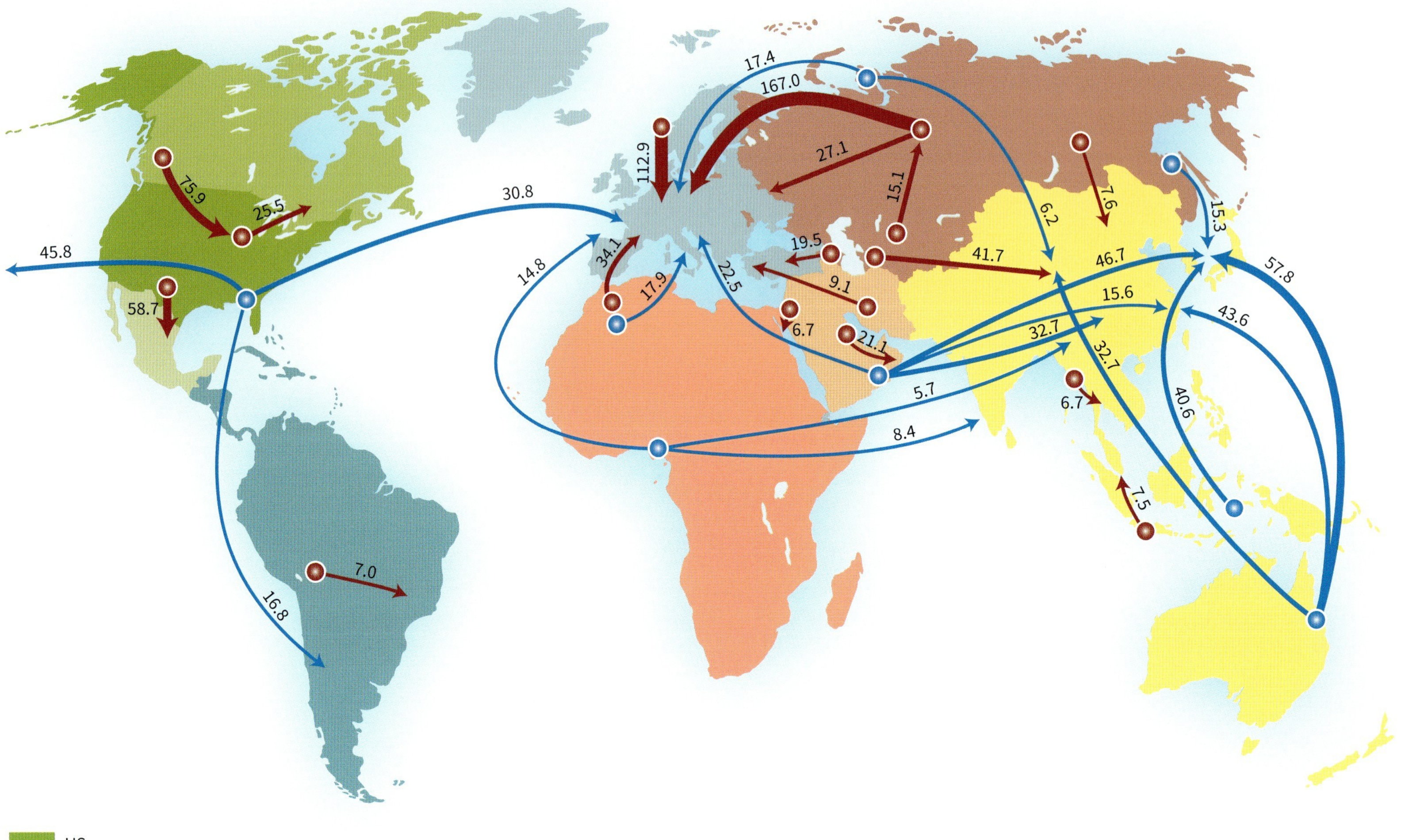

- US
- Canada
- Mexico
- South and Central America
- Europe
- CIS (Commonwealth of Independent States)
- Middle East
- Africa
- Asia Pacific

Billion cubic metres (bcm)
of natural gas, 2021
(1 bcm ≈ 10 TWh)

- < 25
- 25 - 50
- 50 - 100
- > 100

- Pipeline gas
- LNG

The Middle East and Russia are big suppliers of oil and gas, while Europe, Japan, and Asia are big customers. The United States is essentially self-sufficient

Variations in space and time

There is more sun in deserts than at higher latitudes. And the wind blows harder over oceans than over land. There are also considerable time variations in the amounts of solar, wind, and practically all other sustainable energy sources. What do these variations mean in the establishment of a sustainable energy system?

Strong variations

The output of many sustainable energy sources depends on the place on Earth, day/night variations, seasonal variations, and the weather. In the Sahara, a solar panel generates 2-3 times as much energy than it does in (Northwest) Europe; but at night, production at both locations is zero. Wind energy is strongly dependent on the weather. The presence of high- and low-pressure areas determines how hard the wind blows and its direction. The energy from sustainable sources therefore varies greatly in time and space. In this regard, these energy sources differ significantly from fossil energy sources. There are two sustainable sources that are not indirectly produced by solar irradiation. Tidal energy is a product of the gravitational pull of the sun and the moon, and therefore subject to well-defined daily and seasonal cycles. Geothermal energy, or terrestrial heat, is, like fossil energy sources, dependent on the Earth's geological structure, and in principle is characterized by a continuous supply.

Not the same amount of sun everywhere

The output of solar panels is a function of the amount of solar irradiation, the atmospheric conditions, such as cloud cover and air pollution, and factors like the shading and soiling of the panels themselves. The most important factor is the amount of solar irradiation, which depends on the latitude on Earth, the hour of day, and the season. At the equator there are no seasons. The sun rises there every day at 6 in the morning and sets at 6 in the evening. As one moves further north or south, the seasonal variations become greater, and therefore so too do the variations in the number of hours of sun per day. Since the ascending warm air at the equator often results in clouds and frequent rain, the amount of solar energy reaching the Earth's surface is actually the highest at the two tropics. This is where the air descends. Consequently, there is little or no cloud cover, and it practically never rains. At or around the equator there are tropical rainforests; at or around the two tropics there are deserts.

Solar panel robot washer

Average annual solar irradiation [33]

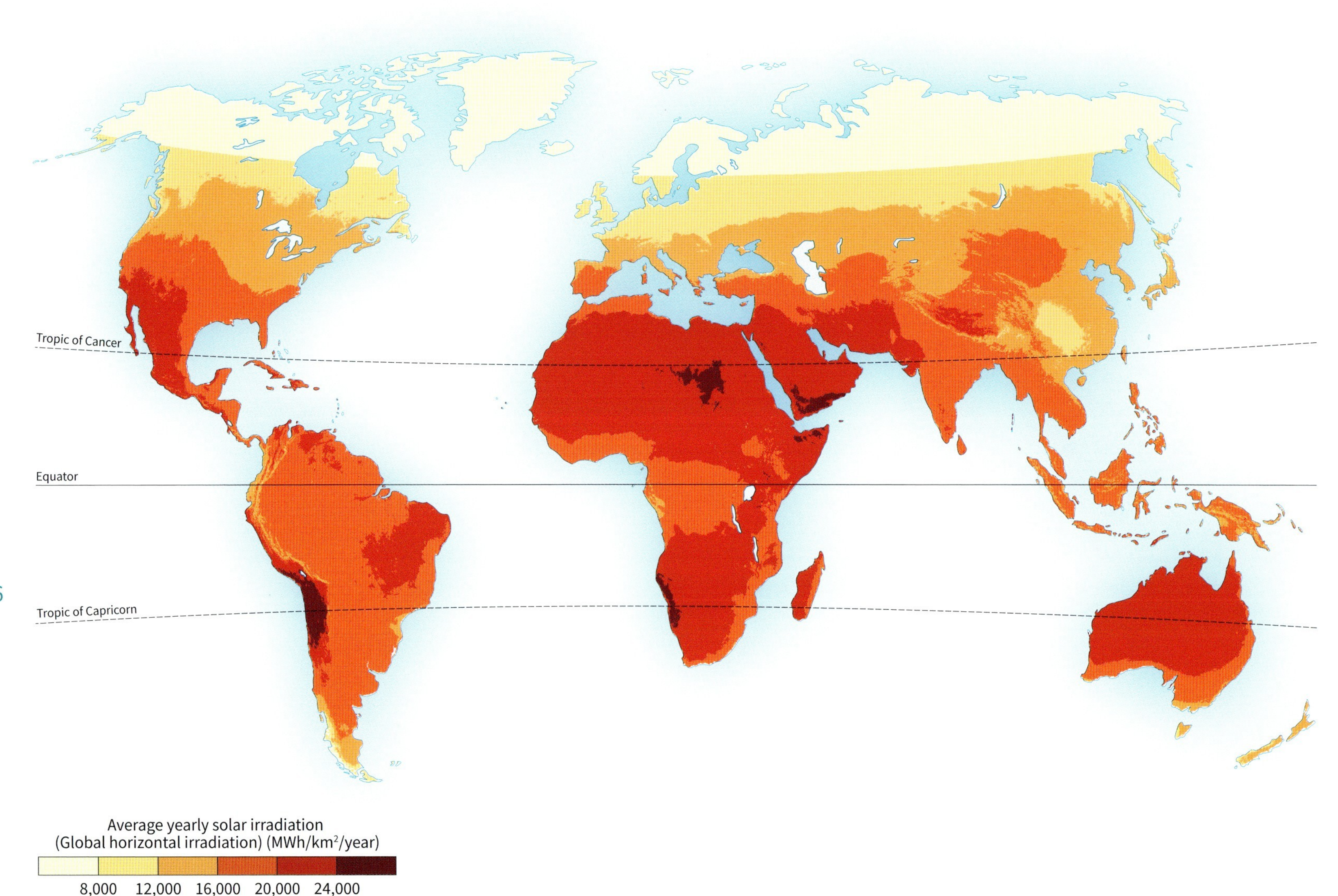

Deserts and oceans are the most suitable places for the large-scale, cheap production of sustainable energy

The amount of solar energy reaching the Earth's surface is the highest in desert areas located around the two tropics. In the southern hemisphere there are desert areas in Chile, Namibia, South-Africa, and Australia; in the northern hemisphere, these areas are found in the south-western part of the United States, in Mexico, the Sahara, and the Gobi Desert in China. The amount of solar energy is two to three times lower in the north of Europe, Japan, and the north-eastern parts of the United States and China.

Mean annual wind speed at 100 m height [34]

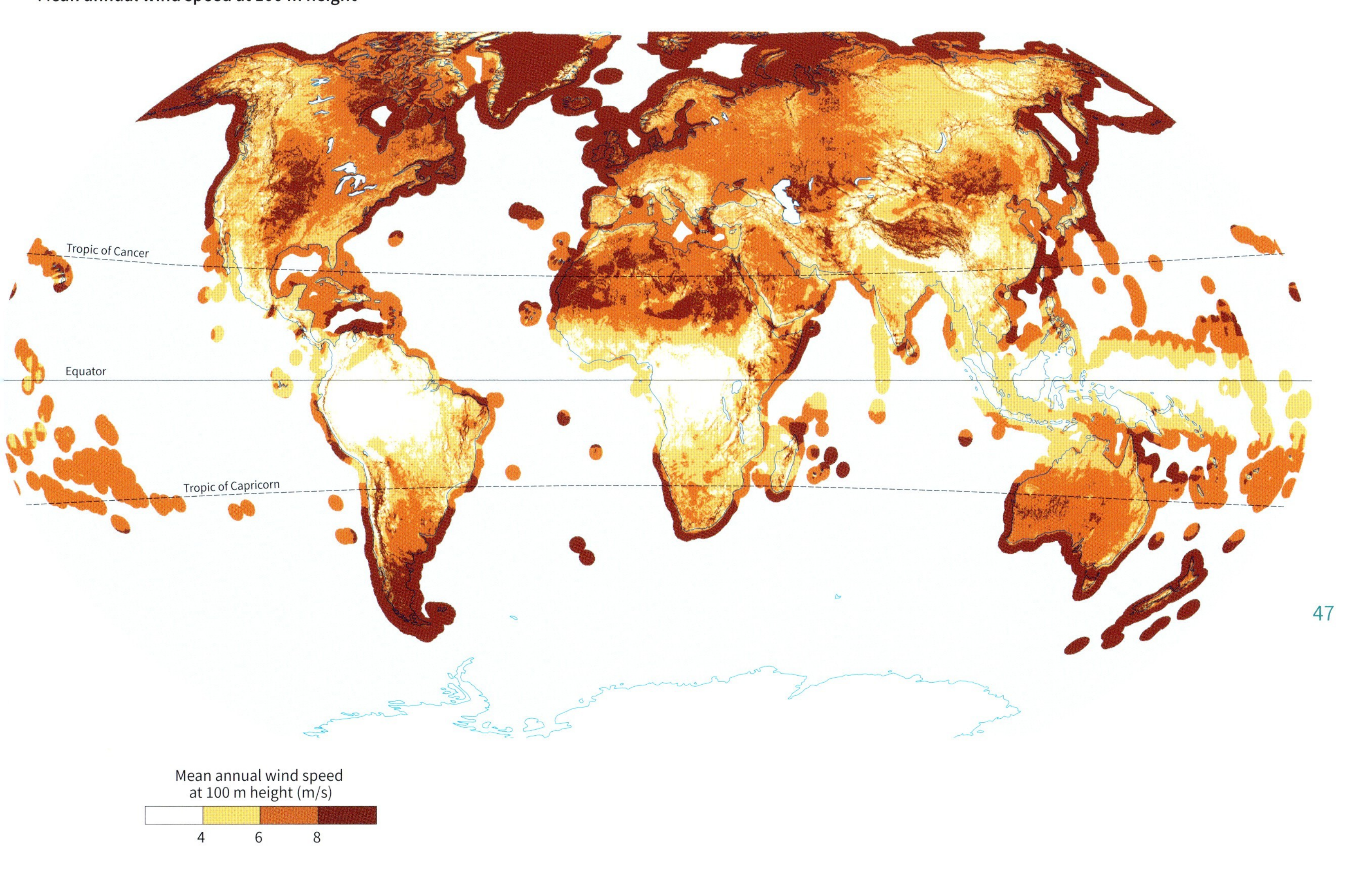

Not the same amount of wind everywhere

The output of wind turbines depends greatly on the wind speed. The amount of wind energy is proportional to the third of the power of the wind speed. Wind, that is, the speed with which air moves, results from the fact that the sun heats the atmosphere unevenly. In places where the sun heats the air greatly, the warm air rises, creating a local low-pressure area. In places where the sun heats the air less, the air is colder and therefore heavier, creating a high-pressure area. Wind is the movement of air from areas of higher air pressure to lower pressure areas. And because the Earth rotates on its axis, the elimination of the pressure differences takes time. High- and low-pressure areas therefore persist for a while. At the equator the air is heated up so much that the air rises perpendicularly. For this reason there is practically no wind there. Because this air descends again at the two tropics, these are areas where strong and regular winds occur, the so-called trade winds. Between the two tropics and the poles the atmosphere is heated irregularly. Temporary high- and low-pressure areas occur in these zones, resulting in wind speeds that can alternate strongly in time. Regional conditions of various sorts can cause wind speeds to strongly deviate from this general picture. In California, for example, a sea breeze arises because a cold sea borders warm land. And in France the mistral is caused by cold air that descends from the mountains.

The wind speed is also determined by the friction of the air with the Earth's surface, with the result that the wind speed at ground-level is zero. The greater the presence of buildings, obstacles, and/or forestation, the greater the friction. And the greater the friction, the more the wind's speed decreases, and the greater its turbulence. The wind speed on the sea is therefore the highest, while it is lowest in cities, where many irregular wind gusts occur.

Around the equator there is practically no wind. In deserts the wind speed is in turn very high in many places, as in the open spaces in the Midwest of the United States, on the steppes of Kazakhstan, and in Patagonia in Argentina. The mean wind speed in many coastal areas is also high. But the highest measured wind speeds have occurred over the oceans, with up to a mean wind speed of 12 meters per second at a height of 100 meters. These are not shown on the map.

Variations in hydro power

The energy output from a hydro power plant is directly proportional to the volume of water that flows through the turbine and to the difference in elevation. Hydro power can therefore only be generated in places where there is a large supply of water (rivers) and a sufficient elevation difference.

River water comes from rainfall or snow melt. Water vapor gets into the air through the evaporation of surface water from oceans, seas, and lakes. If the humid air cools, for instance because the air rises in mountainous areas, then the water condensates creating precipitation in the form of rain or snow. Cold air can hold less water vapor than warm air. A comparable phenomenon occurs above the equator. Humid air there does not rise because of the presence of mountains or hills, but because the solar irradiation has heated it up. Because the air rises, it becomes cool again, with the result that equatorial areas have relatively high levels of precipitation. This has led to the growth of tropical rainforests. The water flowing in the rivers in these rainforests can be used to generate electricity.

The volumes and timing of the flow of rain or melt water depend on the seasons, the temperature, and the weather. Thus, the water reservoirs in Spain are mostly replenished by the rainwater in the autumn and winter period, since little or no rain falls in the summer. Whether hydro power can generate sufficient electricity in the summer therefore depends on the amount of winter rain.

Variations in biomass

Biomass refers to material of plant or animal origin, such as wood, corn, algae, manure, food remains, and plant oil. Biomass grows on the land and in the water thanks to solar energy and water. Because the sun shines and it rains a lot around the equator, biomass production in the area is the highest. The sun also shines often in deserts, but there is little rain. Therefore, not much biomass grows in deserts. Biomass production in other parts of the world lies somewhere between these two extremes. Besides what nature has to offer in the form of biomass, you can also cultivate it for your own needs, as, for instance, in the case of agricultural crops and wood.

Biomass cultivation for energy production can take up space that would otherwise be available for nature or for biomass cultivation for food and materials. When we refer here to the use of biomass cultivation on land as a source of energy, we only mean biomass residues, such as fruit and vegetable waste, manure, sawdust, and sludge. These residues can be used not only as sources of sustainable energy, but also as sources of carbon. In the form of CO_2, CO, or C, carbon is an important feedstock for chemical products and transport fuels. Carbon from biomass residues will in the future provide an alternative to carbon from fossil sources.

The above-mentioned biomass residues are generated primarily on agricultural land, and in industries that process agricultural products into food. Most arable farming and livestock production areas are situated within a reasonable distance from population centers. This means that energy can be generated from biomass residues close to where the energy demand is located [35].

Where is sustainable energy production possible?

The amount of sustainable energy you can produce with solar, wind, hydro, and biomass residues is large enough to provide all of the world's population with energy. Solar and wind energy offer by far the greatest potential. Hydro power can make a good contribution at specific locations in the world. And biomass residues are not only interesting as a source of sustainable energy, but principally as a source of carbon.

The oceans occupy more that 70% of the planet's surface. The land surface consists approximately 20% of barren land and 10% of glaciers. The remaining area, about 70% of the land surface, is habitable in principle. Of this habitable part, half is being used for agriculture, 37% is forest, 11% is shrub, 1% is water, and 1% is built-up.

We can generate solar and wind energy cheaply in areas with high levels of solar irradiation and/or high wind speeds. Large-scale energy generation is best done in places where there is sufficient space, and where the energy production does not compete with other forms of land-use, such as agriculture and nature. Since solar panels occupy a large area, the production of solar energy is difficult to combine with agriculture and nature. But you can effectively integrate solar panels into the built environment, on roofs for instance. The noise nuisance and shadow flicker are among the factors that make the integration of wind turbines difficult into the built environment, but you can effectively combine them with agriculture.

For the above reasons, deserts and oceans are the most suitable places for the large-scale production of sustainable energy [37]. Moreover, in instances where you can produce energy from hydro and geothermal power in an environmentally responsible manner, these are also good, sustainable energy sources. Biomass residues can be used in the vicinity of energy use not only for the production of sustainable energy, but principally as a source of carbon.

Land-use in the world [36]

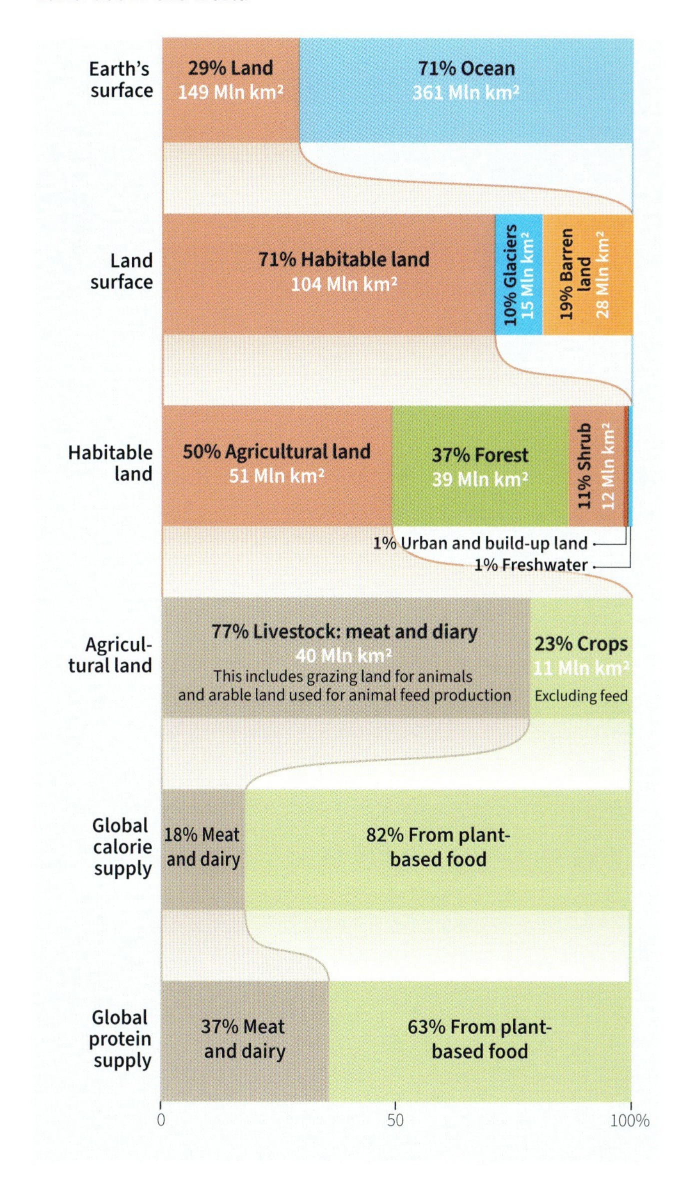

Surpluses and shortfalls of sustainable energy

Where the energy that is used comes from (space dimension) and when it is produced (time dimension) are secondary factors in the analysis of the current, fossil energy system. In the case of a sustainable energy system, however, the dimensions of space and time are of primary importance.

Energy use per km²

The geography of energy use, the energy use per km², has not been previously analyzed. Here we make a first attempt to draw up a world map of energy use per km². Our first step in determining energy use per km² involves multiplying the population density (number of people per km²) by the per capita energy use of a country's population. The energy use for heating and cooling of buildings, mobility, lighting, and power (for machines and appliances) depends to a great degree on the population density, in addition to the weather and the standard of living. There are also spatial variations that this form of calculation does not reflect. For instance, industrial energy use is concentrated in areas where industry is concentrated; and energy use in mining is concentrated in areas rich in mineral resources. The processing of mining products, such as steel production from iron ore, is energy intensive and takes place in areas where energy is cheap. This also applies to basic chemicals. Fossil energy extraction, which consumes energy, occurs in places where this energy is abundant, and where it can be done cheaply.

This map shows clearly that energy use per km² is (very) high on the east coast of the United States, in Europe, the Middle East, India, western China, Japan, and Indonesia. In India and Indonesia this is primarily because of the high population density. In India the average population density is 477 inhabitants per km² [39], and the annual per capita energy use is 6,990 kWh [31]. The average population density in the US is 37 inhabitants per km² and the annual per capita energy use is 76,635 kWh—over 10 times more than in India. However, because almost 13 times more people live on one km² in India than in the US, annual average energy use per km² in India is actually higher than in the US: 2,990 MWh/km²/year versus 2,835 MWh/km²/year. In the US there are big differences between cities and the rural areas; in India the population density is also higher in cities, but the difference between cities and rural areas is smaller.

The map also makes it clear that hardly any energy is used in desert areas, the Himalayas, the Amazon region, and in the northern parts of Canada and Russia. What this map does not show is the energy use on the oceans. Hardly any energy is used there—except by ships and airplanes, but these fill up on land.

Solar energy supply per km²

The supply of sustainable energy is potentially very large. Far larger in fact than the world's energy use, even with a much larger world population and a much higher energy use per head of population.

Now that we have a world map of energy use per km², the next question concerns where the supply of sustainable energy is located, and what factors influence this supply.

Energy use per km² [31], [38]

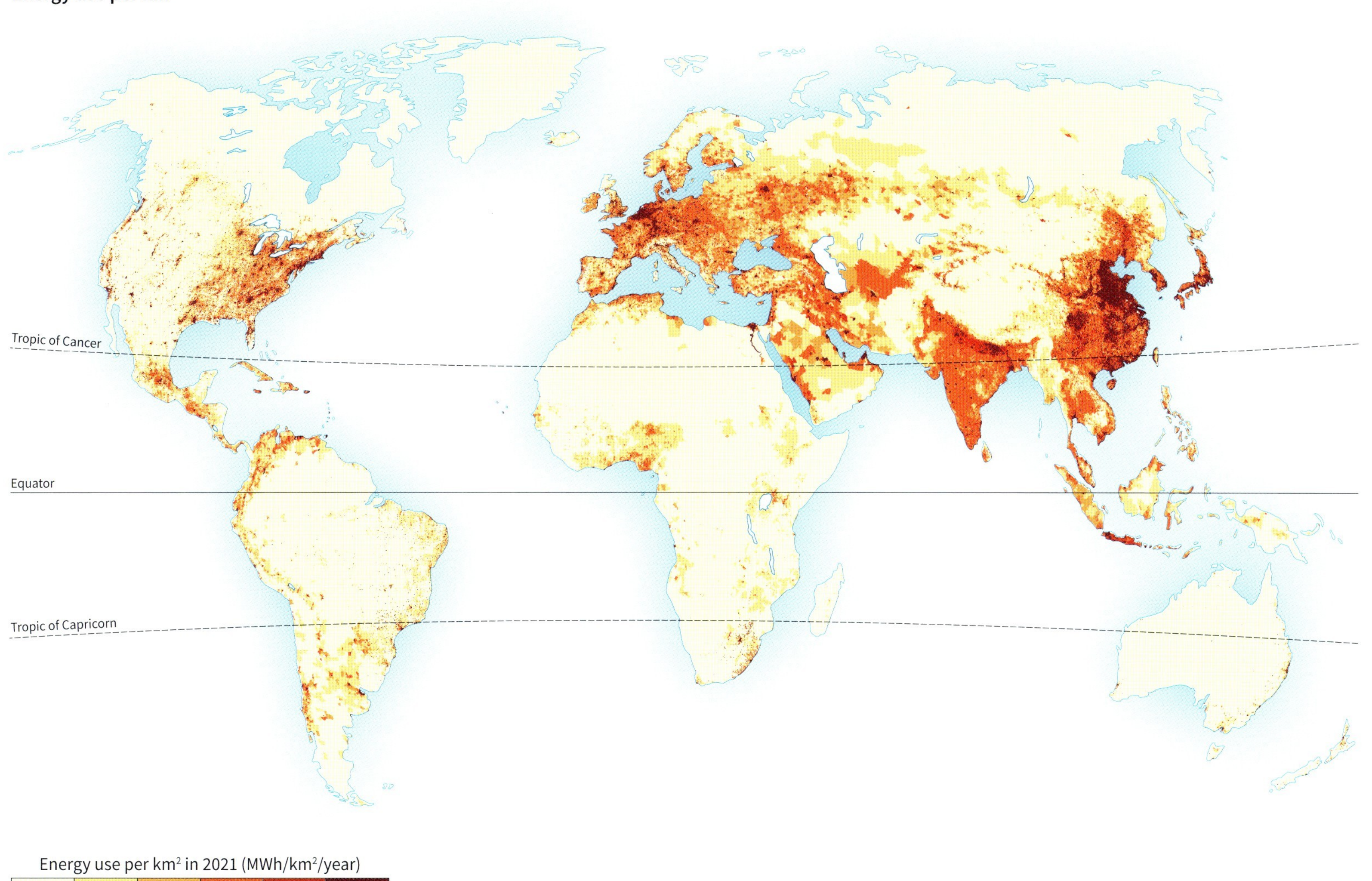

Besides the amount of solar irradiation, for which we have already presented a map, there are further important factors.

In the first place, there are the technical factors, like the efficiency with which you can convert solar energy—for example—into electricity. Secondly, because one row of solar panels can throw a shadow on the next row, there needs to be a certain distance between them.
There are additional spatial factors. It is difficult for instance to install solar panels in mountainous areas; while it is not, or almost not, possible to do so in nature areas, forests, wetlands, and other protected areas. Thirdly, there is the question of competition for scarce space. You can't set up solar panels on agricultural lands or airports. You can however integrate them to a certain extent in the built environment, on industry and recreation sites, and next to transport infrastructure. The number of solar panels you can install in the built environment depends on the population density. Because of this competition for space, the placement of solar panels in densely populated areas will be more costly than in thinly

populated ones. Over time, this might even mean that the generation of sustainable energy in densely populated areas will disappear because imported sustainable energy is cheaper, or because other forms of spatial use receive preference. One example of the latter concerns the competition in cities between the development of green areas for heat-stress reduction and recreation, on the one hand, and solar panels for the generation of sustainable energy on the other.

How we incorporated these factors when making a world map of the supply of solar energy per km^2 is shown in https://www.kwrwater.nl/en/green-energy-for-all/. In making the map, account was therefore taken of the actual available area, by for example excluding agricultural land, airports, nature areas, and other protected areas.

World heat map: surplus and shortage of sun

By subtracting the energy demand from the solar energy supply per km^2, one can produce a world map of areas with surpluses and shortages of solar energy. The world map of the difference between supply and demand we call a "heat map", which, in this case is a heat map for solar energy.

This map shows large-scale shortages in densely populated areas that are surrounded by farmland, as in India, but also in Western Europe, and the Northeastern US. There is often a surplus however in areas located farther away from population centers. Shortages are not only present in industrialized countries with high per capita energy use, but also in countries with high population densities, like India, Indonesia, and the Philippines. If the per capita energy use increases further in these countries, then the shortages rise even more.

Even though the oceans are not colored in on this map, it is nonetheless on the oceans, especially far from the coast, where there is a considerable surplus of solar energy. The supply is limited closer to land by a variety of factors, such as the use of waters for fishing, nature, and shipping.

In places with shortages, sustainable energy can be generated locally, with solar panels on roofs of buildings, for example. But the amount of energy generated in this way is not sufficient to cover all of the energy use. Consequently, there is a shortage in these areas. It is also conceivable, and perhaps even necessary, that in the future our consumption of animal protein will decline, which would free up stock-raising areas. Part of these areas could be used to increase the supply of sustainable energy. And maybe, in certain countries or regions, the competition for scarce space could actually be settled in favor of sustainable energy production. But, on the other hand, the world's population will continue to grow and, furthermore, increased prosperity will stimulate greater energy use—though admittedly more slowly, thanks to greater use-efficiency.

Besides the amount of sun, other factors are important for the production of solar energy: technical factors, space availability, and competition for scarce space

Our conclusion is therefore that areas with a high population density and/or a high energy use per capita will necessarily import sustainable energy. And areas with a low population density and good sustainable energy sources will then export sustainable energy. This means that in a global sustainable energy system the transport and storage of sustainable energy will be important.

Solar energy heat map: surpluses and shortages per km² [31], [38], [40], [41], [42], [43]

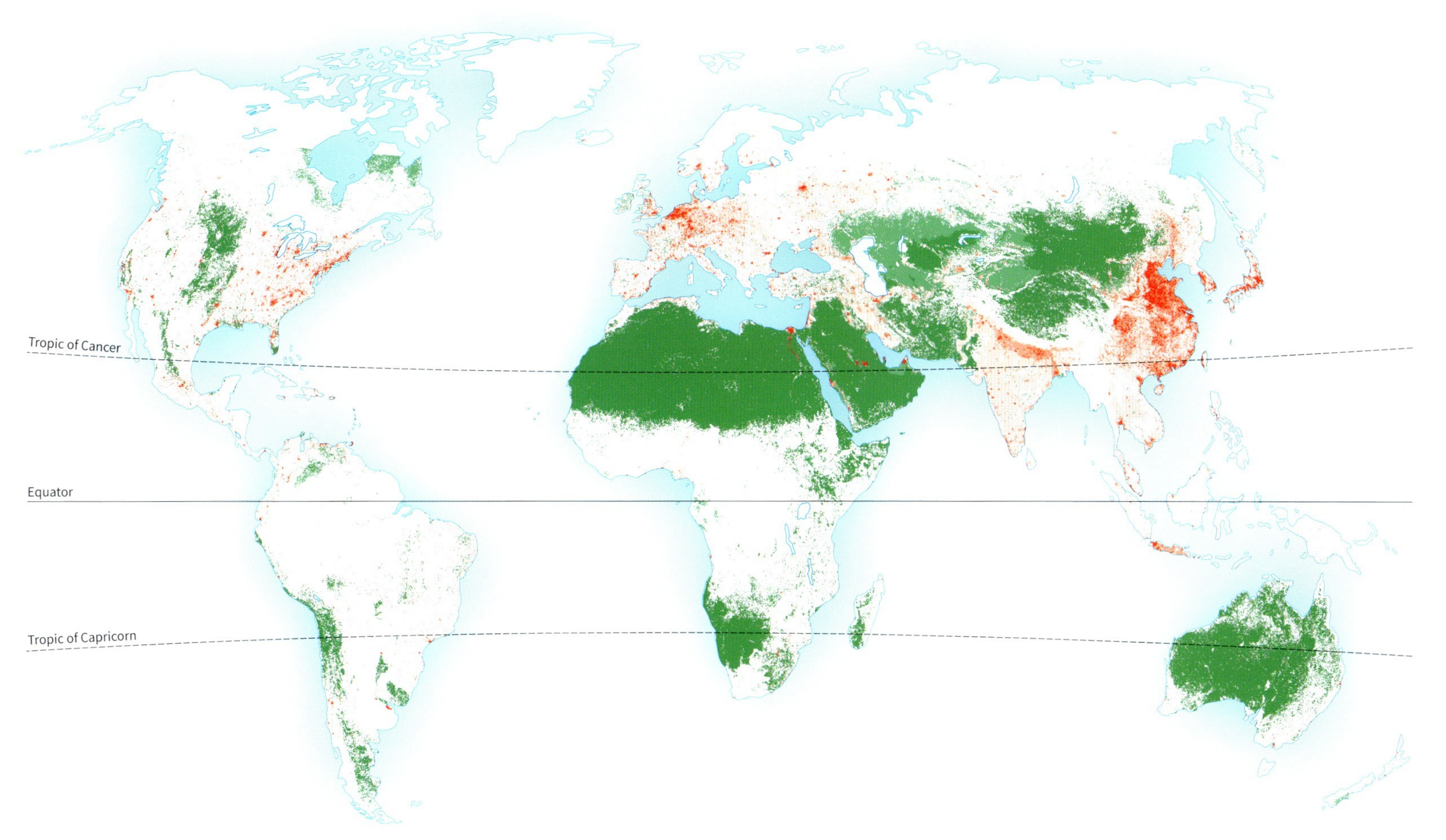

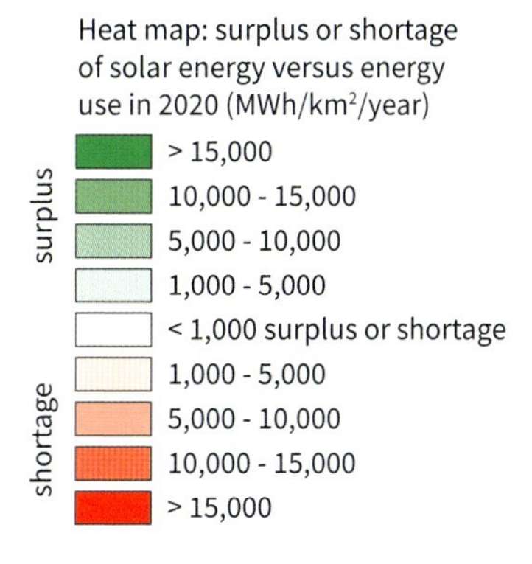

In a global sustainable energy system transport and storage of sustainable energy will be important

How do we transport energy?

In a fossil energy system, you transport mainly coal, crude oil, and natural gas, and the transport costs do not represent a big share of the total costs. In a sustainable energy system things are very different.

In the current, fossil energy system, there is a worldwide transport by ship of energy in the form of coal (solid substance), oil, or liquefied natural gas (LNG) (liquid). Intercontinental, long-distance transport of energy, in the form of oil (liquid) or natural gas (gas) for instance, can also be done using pipelines. The transport costs in today's fossil energy system represent at most 10% of the total costs [44].

In a future, sustainable energy system, the percentage share of transport costs will be bigger for two reasons. First of all, it is very probable that more (sustainable) energy will be transported over larger distances than is the case in a fossil energy system. Secondly, solar panels and wind turbines, in the first instance, produce electricity, an energy carrier that is much more expensive to transport than coal, oil, or gas.

Electricity transport via high-voltage lines

Earlier we saw that the future, sustainable energy system will primarily draw on the sun and wind as energy sources. At good locations, solar energy can be produced for € 0.01/kWh [45], and electricity from the wind for € 0.02/kWh [46], [47]. In the future these production costs will drop further. The cost price of sustainable energy production is therefore lower than that of fossil electricity production. It is also lower than the cost price of fossil sources like natural gas, crude oil, or coal—certainly if a CO_2 tax is included in the calculation. However, these low production costs for solar and wind apply only to locations with lots of solar irradiation or high wind speeds, as well as extensive cheap available space. These are not the same locations as the locations with high demand for energy. The question is therefore how much does sustainable energy cost if one delivers it at the desired location and time. This cost price includes not only the production costs, but also the costs for transport and storage.

Transport by ship, pipeline, or cable

Energy in the form of a fluid or of a solid has a high energy density per volume unit (in kWh/liter). It is also easy to transport by ship. You don't need to convert coal and crude oil into another form for transport purposes. Since the energy density (in kWh/l) of natural gas is a lot lower, you have to convert it into LNG before you can transport it cheaply by ship. Despite the extra cost of this conversion, you can deliver the natural gas to customers at a lower cost than if you had transported the natural gas as a gas by ship. Fifty years ago, when the possibility of producing LNG from natural gas still didn't exist, natural gas was converted into another liquid: methanol. This is the reason that there is a large methanol production capacity in areas where natural gas is extracted at a great distance from the demand—for instance, in Trinidad and Tobago, New Zealand, and Southern Chile.

Can you transport electricity by ship? Not in the same way as crude oil or natural gas, because you can't store electrons in a tank. You can however store electricity in batteries and transport these by ship. But the energy density of batteries (in kWh/l and kWh/kg) is so low that the transport of electricity by ship is very costly. Nor can you get electrons to flow through a hollow pipeline, but you can do it with a conducting cable. This approach entails a loss of electrons caused by the resistance over the course of the transmission. The electricity loss is equivalent to the square of the current. By transporting the electricity under high voltage, you can deliver the same power with less current, and in this way limit the energy loss. The conversion to a higher voltage is done using a transformer, which costs money and itself involves a loss. Nonetheless, the costs for the transport of electricity over long distances are lower when it is done using high-voltage lines.

Energy density per volume and weight for different energy carriers [1], [48], [49], [50], [51], [52]

	Energy density by weight kWh/kg (HHV)	Energy density by volume kWh/liter (HHV)
Hard coal	6.6-8.6	9.9-13.0
Wood pellets	4.9	3.1
Crude oil	12.4-13.0	11.2-11.7
Gasoline	13.2	9.7
Diesel	12.8	10.6
Aviation fuel	12.8	10.5
Methane (natural gas), gaseous	15.4	0.011
Methane (natural gas), liquid -162 °C	15.4	6.6
Hydrogen, gaseous	39.4	0.0035
Hydrogen, liquid -253 °C	39.4	2.8
Ammonia, liquid -33 °C	6.25	4.35
Electricity, battery storage (lead acid-lithium range)	0.04-0.40	0.03-0.27
Hot water, 90 °C	0.1	0.1

Transport costs and transport capacity for different energy carriers [44], [53], [54]

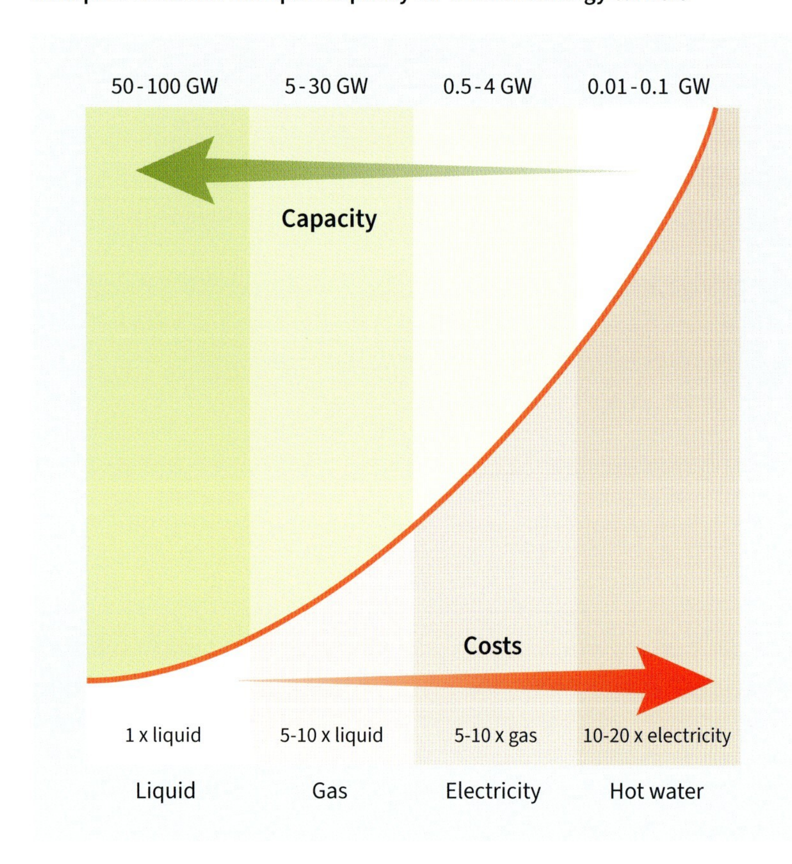

Apart from electrons (electricity) and molecules (liquid or gaseous energy), you can also transport heat or cold in the form of water. Hot or cold water can be transported via pipelines, for instance for use in the heating or cooling of buildings. But you can also transport steam—gaseous water—under pressure through pipelines, to deliver process heat in industry for example. The energy density (kWh/l and kWh/kg) of hot and cold water is low when compared to other energy sources and carriers; furthermore, large, well-insulated heat pipelines are needed for the transport of relatively small volumes of energy.

Energy transport costs and capacity

The cost of the transport of energy depends greatly on the energy density per volume (kWh/l) and, to a lesser degree, on the energy density per weight (kWh/kg) of the energy carrier. For the transport costs the following applies: the greater the energy density, the lower the costs. The transport of a liquid through a pipeline is the least costly. Then comes the transport of a gas through a pipeline, which costs 5 to 10 times more. Transport of electricity by a power line is roughly 5 to 10 times more expensive than the transport of gas through a pipeline. The costliest option is the transport of heat or cold water through an insulated pipe: 10 to 20 times more expensive than the transport of electricity using a power line [44], [53], [54].

The transport of liquid or gas molecules through a hollow pipe does not (in principle) involve the loss of any molecules. However, to transport a large volume of gaseous energy through a pipe you need to compress that gas in order to increase the energy density per volume. For this compression, you need energy, usually electricity, so that there is effectively an indirect energy loss. The compressors used, also involve an investment. Additionally, the friction of the pipewall during the transport causes a drop in pressure, and you can use compressors to regularly bring the pressure back to the desired level. The compression does nevertheless lead to lower costs for the transport of natural gas or hydrogen gas over long distances.

Gas transport is 5-10 times more costly that liquid transport; electricity transport is 5-10 times more costly than gas transport; hot-water transport is 10-20 times more costly than electricity transport

For the transport capacity the following applies: the greater the energy density, the larger the capacity. A pipeline to transport liquid has a capacity of 50-100 GW. A gas transport pipeline has a capacity of 5-30 GW. For example, one “Nord Stream” pipeline—for the transport of Russian natural gas through the Baltic Sea to Germany—had a capacity of 35 GW. Electricity cables generally have a capacity of 0.5-4 GW, although a cable has been produced in China with the capacity of 12 GW. Heat pipelines for the transport of hot or cold water have the lowest capacity: 10-100 MW, or 0.01-0.1 GW.

Energy transport in an energy system

The transport costs are key in determining the logistics of an energy system. The energy transport system can be characterized as follows.

- Worldwide transport is done by ship. Energy is transported in the form of a liquid or solid. In order to transport gas by ship, it must first be converted into a liquid.
- For distances up to about 5,000 km, energy is transported as a liquid or gas via pipelines. For the cost-effective transport of electricity over thousands of kilometers, it is first converted into a gas or liquid.
- Up to about 1,000 km, electricity can be transported at a reasonable cost by electricity cable.
- Up to about 50 km, heat and cold can be transported at a reasonable cost via heat pipelines.

In the current, fossil energy system, it is primarily hydrocarbon molecules (coal, crude oil, and natural gas) that are transported by pipeline. This is a simple matter. Apart from the compression of natural gas into LNG, no conversions are necessary. In this system, the electric power plants that run on crude oil, natural gas, or coal, are located near the electricity demand. This is by far the cheapest solution, because the transport of electricity is far more expensive than the transport of coal, crude oil, and natural gas.
In a sustainable energy system, electricity can of course be produced a great deal more cheaply, in places with lots of solar irradiation and/or high wind speeds, and where there is plenty of inexpensive space available. These areas are often located at great distances from the energy demand. To make the cost-effective transport of this electricity possible, you first convert it into a gas, hydrogen gas. And to transport this gas by ship all over the world, you need to convert it into a liquid, such as liquid hydrogen.
One alternative is to make ammonia by binding hydrogen gas to nitrogen from the air; ammonia becomes liquid at -33 °C. Hydrogen can also be bound to a liquid organic hydrogen carrier (LOHC), an oil-like liquid which can be transported in today's tankers.

At equal energy volumes, the total transport costs in a sustainable energy system are far higher than in a fossil energy system. This is the result of the extra costs for the required conversion of electricity into a gas and liquid, which allows you to achieve cheaper transport costs.
The share of transport costs in the total costs can no longer be neglected, as in the case of a fossil energy system. This is why one can't compare fossil and sustainable energy systems solely on the basis of their production costs. The transport costs also belong in such comparisons, for instance as part of an energy scenario analysis.
We estimate that the transport and storage costs could together actually account for about 50% of the total costs in a sustainable energy system.

Energy transport method and range

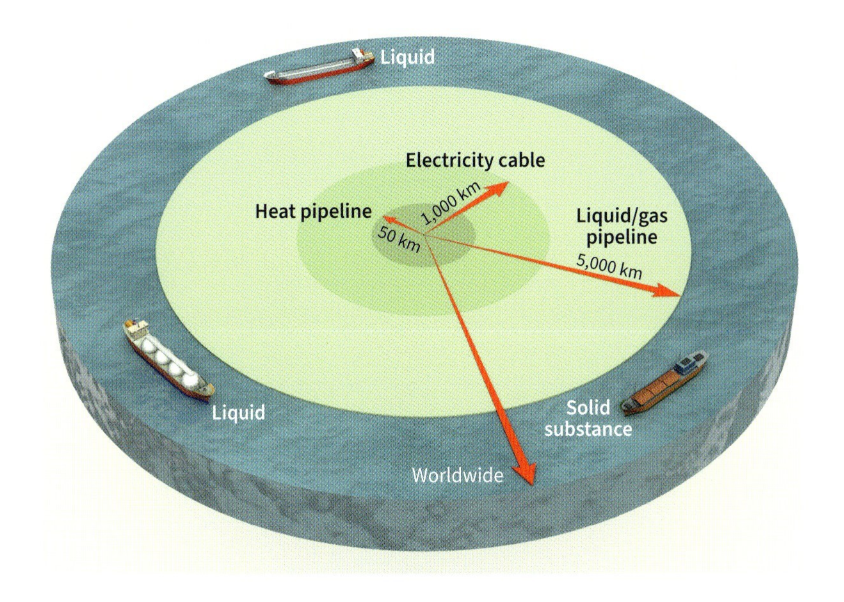

How do we store energy?

Energy storage is needed so that energy demand can be met at any time. The current, fossil energy system, disposes of energy storage to respond to variations in energy demand. A sustainable energy system faces variations not only in energy demand but also in energy supply. It therefore needs more energy storage.

Energy storage in a fossil energy system

The extraction of fossil energy provides a more or less continuous stream of energy. Every day, comparable amounts of coal, crude oil, and natural gas are extracted. You can easily transport such a continuous stream of oil or gas via pipelines. For ship transport, one needs to have storage possibilities in the ports, so that the ships can be quickly loaded and unloaded. Storage of fossil energy is easy and cheap: coal can be stored in the open air, oil in oil tanks, and gas in empty gas fields, salt caverns, suitable aquifers, and rock domes.

Energy storage is needed mainly as a means of matching supply and demand. Natural gas requires quite a large storage volume. In countries where buildings have to be heated in the winter, the demand for natural gas has a seasonal pattern: high demand in the winter (to heat buildings), and low demand in the summer. Because the supply of natural gas is continuous, gas is stored in the summer for the winter. Moreover, many countries use gas-fired power plants to maintain the supply and demand of electricity in balance. Coal and nuclear power plants supply a steady flow of electricity, the baseload. Gas-fired power plants, gas turbines, and gas engines, meet the demand peaks through flexible production. Gas-fired power plants can be effectively adjusted: you can get them quickly and easily to produce more or less. They are also now used to compensate the irregular supply of solar and wind electricity. In this way they provide flexibility to the electricity system, so that supply and demand are always in balance. On islands with a small electricity system, diesel engines are the primary means of ensuring the flexibility in the electricity system.

Spherical tanks for liquid-hydrogen storage

Gas-fired power plants, gas turbines, and gas engines can be effectively adjusted; you can get them easily to produce more or less, and in this way they provide flexibility to the electricity system

In the European Union as a whole, 2 to 4 times more gas is used in the winter months than in the summer months; but in Northern Europe the figure is actually 5 to 10 times more. To satisfy the total demand for gas in the winter, gas storage facilities need to be stocked for the winter. To this end, the EU has a storage capacity that amounts to about 25% of its total annual gas use [18], [55].

Energy storage is also needed for the strategic reserve. In a fossil energy system the costs of storing energy are low, representing only a small percentage of the total costs. This is because the investment costs for the storage of coal, oil, and gas are low. The investment costs for gas storage in empty salt caverns, for example, are about € 0.20-0.40 per kWh of storage capacity [56]. The costs of gas storage in empty gas fields are lower by a few more factors. Investment costs for oil storage in tanks are in the order of € 0.01 per kWh [57]. The costs for coal storage in the open air are much lower still.

Forms of energy storage

Energy can be stored in a variety of ways, using different storage technologies: in chemical energy (coal, wood, oil, natural gas, hydrogen, ammonia, methanol, etc.); in electrochemical energy (supercapacitor, batteries); in mechanical energy (flywheel, pumped hydro power, compressed air); in heat/cold (hot water, phase-change material (PCM), thermo-chemical storage); and in a few other forms [58].

Energy storage in an energy system can be characterized by three metrics: energy content, power, and charge/discharge timescale.

- **Energy content in kWh.** This metric is determined to a great extent by the energy density (kWh/l). The storage method or technology moreover determines how much volume and weight you need to add for the storage itself. For oil, all you need is a tank with a relatively thin wall; for gas under pressure, you need a tank with a thicker wall, which means its volume is greater and it is heavier.
- **Power in kW.** This quantity represents the amount of energy you can draw from the storage per unit of time. For a battery, this is determined by the power electronics around the battery, while for a tank it is a function of the pump capacity. The relationship between the storage volume and the storage power indicates how long the storage can supply energy when it is run at full power until depletion. For stationary batteries, a period of four hours (or a multiple of four hours) is often realized.
- **Charge/discharge timescale.** This is the time that you can store the energy without significant losses. Flywheels lose a few percent of their energy after an hour. The loss in the case of heat storage in tanks depends on the insulation, but it is generally on a timescale of days to weeks. Lithium-ion batteries lose 1-5% of their energy every month [59]. When you store molecules in a tank, they generally undergo no losses over time. The storage can last months, even years, with no energy loss. True, during the storage of liquid gas or hydrogen, depending on the insulation, some of the liquid will evaporate, or “boil off”, and return to a gaseous state. The boil-off rate of a stored liquid gas, such as liquid hydrogen, is 0.1-2% per day, depending on the insulation [60]. But since no molecules are lost, this is, strictly speaking, not an energy loss. Indeed, the boil-off, which is still very cold, can be re-liquefied using very little energy.

Energy content, power, and charge/discharge timescale for different energy storage systems (logarithmic scale) [61]

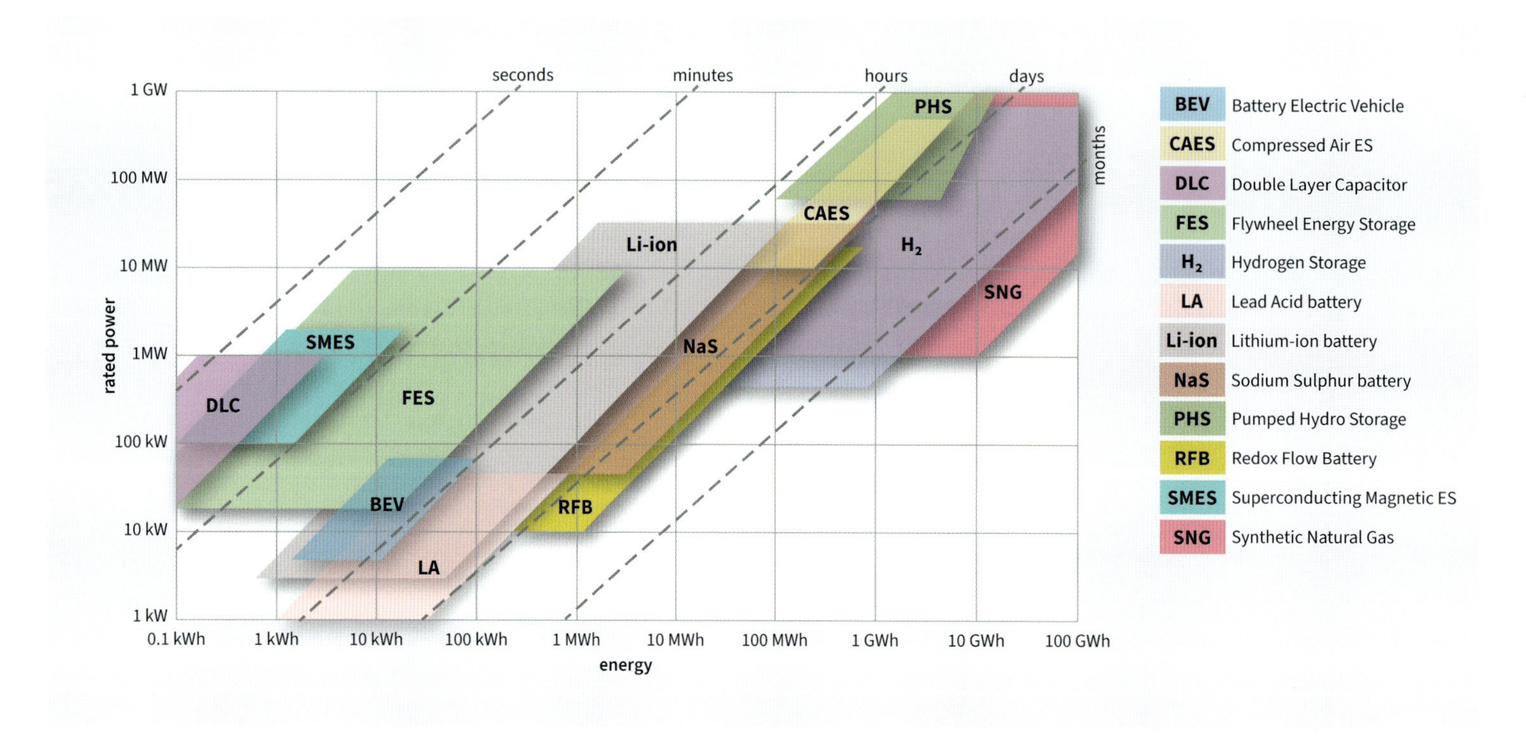

Energy content, power, and charge/discharge timescale for different energy storage systems (linear scale) [61]

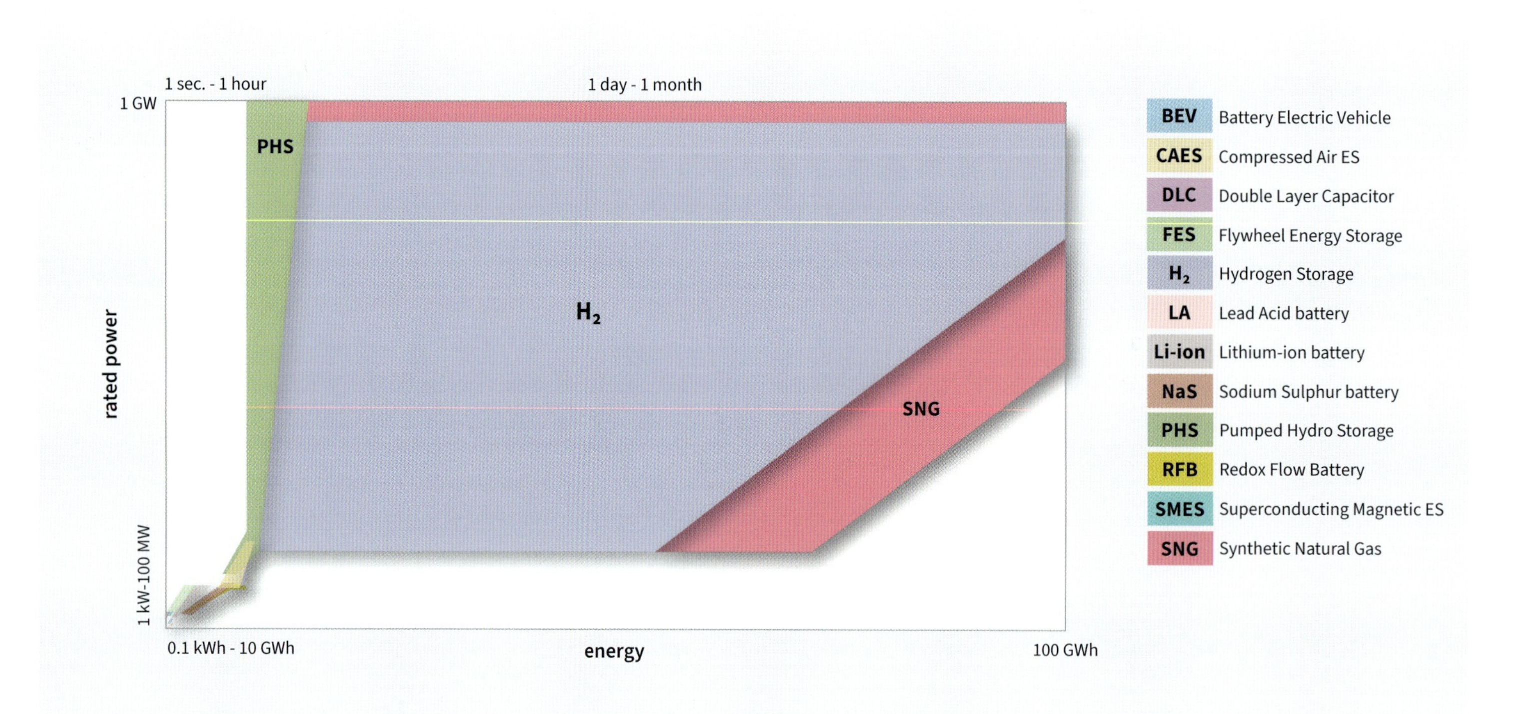

The different energy storage systems are often represented graphically with double logarithmic axes. This presents a misleading picture: it makes it look as though the energy content and the power of molecule storage are only a couple of times larger than those of battery storage. But this is not the case. If you use a graphic with linear axes to compare the same energy storage systems, it then appears that only chemical energy (molecules) can be stored on a large scale over longer periods. In a sustainable energy system this concerns energy in the form of hydrogen or hydrogen derived molecules and synthetic or biofuels.

Costs of energy storage

The investment costs for an energy storage system are, as in the case of transport costs, strongly determined by the energy density per volume (kWh/l).

The investment costs involved in the energy storage of a liquid, such as oil, in a tank under atmospheric pressure, are very low. Hydrogen storage in salt caverns is more costly by a factor of 50-100. The storage of electricity in pumped-up water is more costly, by a factor of 100-200 compared to hydrogen storage in salt caverns. Storing electricity in batteries is up to 4 times more costly than in pumped-up water. And the storage of electricity in super-capacitors is extremely expensive, and is done not with the objective of storing the electricity, but of being able to absorb electricity fluctuations very rapidly.

The storage of electricity is roughly 100 times more costly than the storage of gas molecules; storage of gas molecules is 10-100 times more costly than the storage of liquid molecules

When sorted according to transport costs, each consecutive energy carrier is always more costly than the preceding one by a factor of about 10. When it comes to storage, the differences are far greater: the storage of each consecutive energy carrier is always roughly more costly by a factor of 100: the storage of electricity is 100 times more costly than that of gas molecules, the storage of gas molecules is 10-100 times more costly than that of liquid molecules. But there is an exception: the storage of hot water in subsurface aquifers is less costly than the storage of electricity in batteries.

Investment costs per kWh energy content for different storage systems; sources: [56], [57], [62], [63], [64], [65]

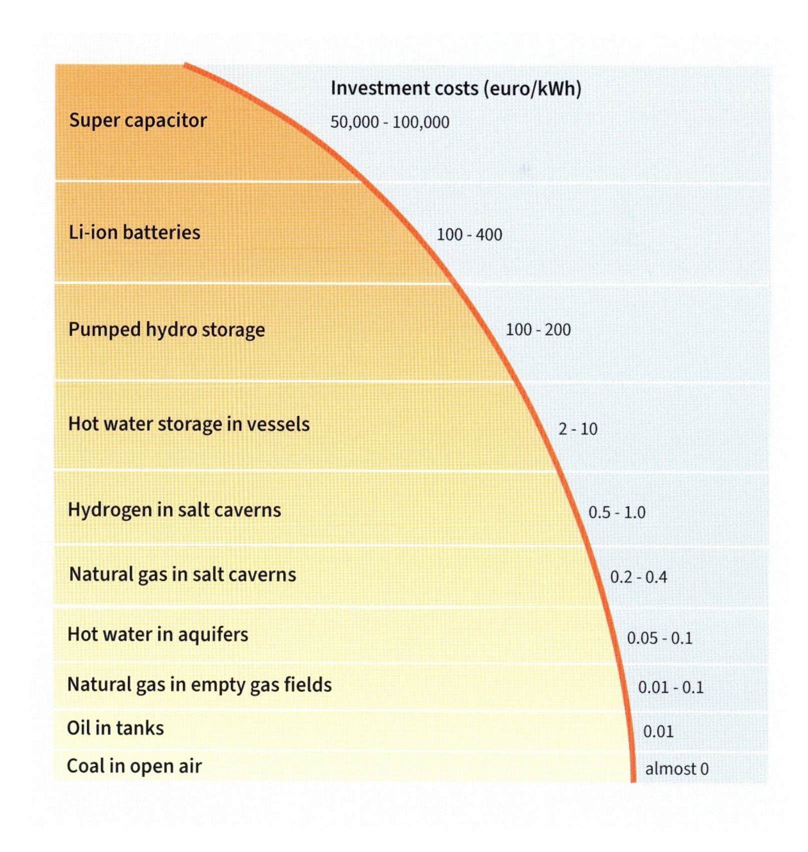

Hydrogen storage in ice crystals

In search of the "holy grail"

We can store hydrogen for lengthy periods, in large volumes, and at a low cost in underground salt caverns and empty gas fields. If a place doesn't have such underground structures, then you store the hydrogen above-ground. This can be done in tanks, or by binding hydrogen to materials or liquids. This is not only more difficult, but also more expensive. For this reason, a lot of research is investigating other possibilities for hydrogen storage—in search of the "holy grail" of sustainable energy storage.

Flemish scientists found inspiration in a natural form of gas storage [66]. They investigated how methane is trapped in ice crystals in permafrost areas (with a permanently frozen ground), and deep in the oceans. Methane actually represents more than 10 percent of the weight of these crystals [67]. This "natural" storage arose over the course of a million years. These scientists are taking the first successful steps toward the storage of hydrogen in ice crystals, employing the same "natural" principle, but of course a lot faster. Practical application will call for much more research, but, who knows, maybe this will eventually lead to a cheap form of hydrogen storage? And all thanks to Mother Nature.

A methane clathrate: methane trapped in ice crystals on the ocean floor [68]

You can't compare a fossil and a sustainable energy system only on the basis of production costs, you must also include the transport and storage costs

Energy storage needs in a fossil and in a sustainable energy system

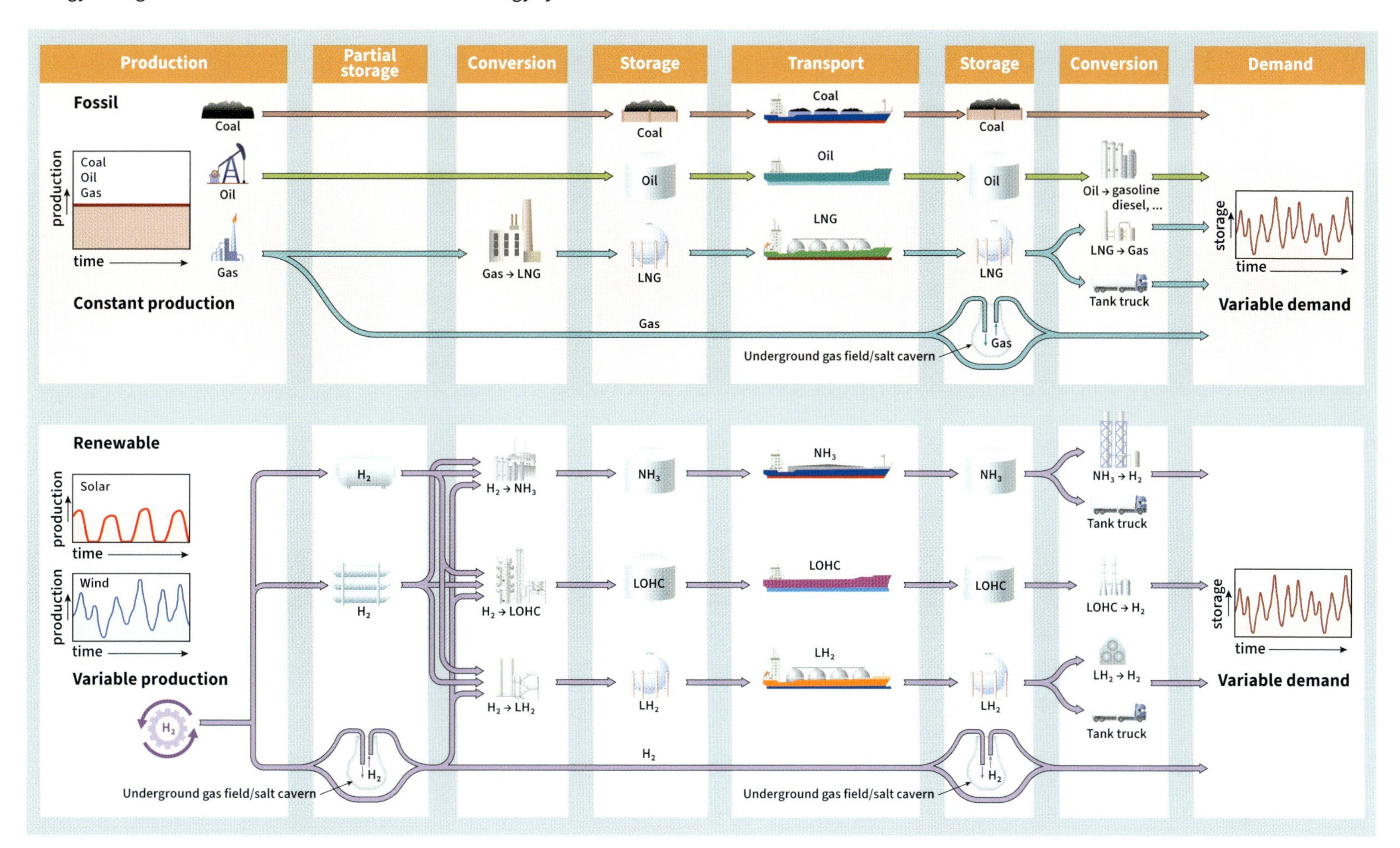

Energy storage in a sustainable energy system

A sustainable energy system needs more energy storage than does a fossil energy system, because, besides the energy demand, the energy supply (sun and wind) is also variable and non-adjustable. Furthermore, one of the two sustainable energy carriers, electricity, is not as easy or cheap to store as coal, crude oil, and natural gas. The question is then: How can one establish a sustainable energy system without making use of fossil sources, which you can use and store easily, cheaply, flexibly, and adjustably? The simple answer is that there needs to be storage in two places in the system: at the supply and at the demand. At, or in the vicinity, of a large-scale production site, storage possibilities are needed so as to transform the fluctuating supply into a continuous supply. To this end, electricity is converted into a gas, hydrogen, which can be stored on a large scale relatively cheaply. The gas can be transported in baseload through a pipeline. Or can then be reconverted into a liquid, such as liquid hydrogen or ammonia, for transport by ship. After the transport stage, the energy system is organized in the same way as the current, fossil energy system: hydrogen gas, ammonia, liquid hydrogen, or otherwise bound hydrogen are stored underground or in tanks, so that you can use them in a flexible and adjustable manner in meeting energy demand at all times.

What does 1 kWh solar energy from Morocco cost in Germany?

If we want to realize large-scale, sustainable energy production systems, then we not only look at the production costs, but at all costs: production, storage, transport, distribution, and end-conversion. We will have to think and act bigger, both when it comes to sustainable energy production and to energy transport over long distances.

How does solar energy from Morocco get to Germany?

We'll discuss an example of solar energy produced in Morocco and delivered in Germany. We first examine what the costs are for the baseload hydrogen supply to Germany of solar energy produced in Morocco. This concerns the total costs, that is, the costs of production, storage, and transport of hydrogen. To begin with, we select an area where you can generate sustainable energy cheaply with solar and wind, and where hydrogen can be stored underground at little cost. There is the possibility of making salt caverns (for the storage of hydrogen) in an area with salt structures off the coast of Morocco, and on land in the Essaouira Basin. More detailed geological research should determine whether this is actually feasible. An interesting site for the generation of solar and wind energy is located to the north of Agadir and extends up to Essaouira. It is therefore a good area to use large-scale generated solar energy for the production of hydrogen. By storing hydrogen in a salt cavern, we can transform the day/night pattern of the hydrogen production from solar energy into a continuous supply of hydrogen. This hydrogen can be transported by pipeline from the port of Jorf Lasfar across the sea to Spain, where the pipeline connects to the European Hydrogen Backbone infrastructure, which runs through France to Germany. This ensures that the hydrogen baseload is delivered to Germany.

Supplying baseload hydrogen to Germany

Suppose that at this place in Morocco we build a 100 GW solar energy system with a surface of 1,800 km^2, with which we produce 4 million tonnes of hydrogen annually (ca. 11,000 tonnes per day). In this case, a single salt cavern, with a storage capacity of about 6,000 tonnes of hydrogen, would be sufficient to provide a continuous hydrogen supply, because day-night supply variations will be the main concern. By means of a 48-inch pipeline, with a capacity of 20 GW (HHV), this hydrogen can be transported over a distance of 750 km to Spain. It can then reach Germany via the European Hydrogen Backbone infrastructure; the total transport distance would be approximately 3,000 km.

Hydrogen, electricity, and heat from Morocco

Hydrogen produced with solar energy in Morocco and delivered as baseload in Germany costs € 1.40-2.00 per kg. This is equivalent to 3.5-5.0 eurocents per kWh. We can use this hydrogen to make electricity again, whenever we like. We will now estimate the costs of the electricity produced in Germany with Moroccan hydrogen. For this, we transport the hydrogen through a distribution network, so that it can be reconverted into electricity by a stationary fuel cell (with an efficiency of 60%) in a residential neighborhood. When we calculate all the costs involved, the result is a price of 6.5-9 eurocents per kWh. And at this price the electricity can be supplied at any time, day and night, summer and winter.

Cost price of hydrogen from Morocco that is delivered as baseload in Germany

Baseload solar hydrogen from Morocco to Germany		Levelized cost of hydrogen LCoH €/kg H_2
	Assumptions	
Solar hydrogen production	Costs solar electricity = € 0.01 /kWh Full-load hours = 2,000 hours/year Electrolyzer efficiency = 50 kWh/kg H_2 100 GW Solar energy = 4 million tonnes H_2 Required surface = 1,800 km^2	1.0-1.5 [18]
Storage in salt cavern	From fluctuating production to baseload; daily cycle 1 salt cavern of 6,000 tonnes H_2 needed	0.1-0.2 [69]
Transport via pipeline	Pipeline capacity = 20 GW Full-load hours = 8,000 hours/year Pipeline to Spain = 750 km Total transport distance Spain-Germany = 3,000 km	0.3 [70]
Hydrogen as baseload delivered in Germany		**€ 1.40-2.00 /kg H_2** = € 0.035-0.050 /kWh H_2(HHV)

Electricity made from hydrogen which is produced with solar energy in Morocco costs 7.5-10.0 euro-cents per kWh, with an added half kWh of free heat

We can also deliver the hydrogen right to the homes and buildings, and use a fuel cell there to make electricity and heat. In Japan fuel cell systems of this type are already commercially available [71]. Naturally, we only get the fuel cell to supply electricity when the solar panels output is insufficient, that is, principally at night and also during the day in the winter. The heat (60-80 °C) that the fuel cell produces simultaneously can be used at night to fill a boiler vessel with hot tapwater, and also to heat the home in the winter.

Because the fuel cell in this scenario does not supply electricity and heat continuously, hydrogen storage is also needed in Germany to meet the variable demand. Moreover, a small-scale fuel-cell system with a boiler vessel is more expensive than a stationary, larger fuel cell for an entire neighborhood. When one adds up all the costs, the estimated electricity price comes to 7.5-10.0 euro-cents per kWh, and you also get extra free heat of more than a half kWh.

There are other benefits besides the free heat. For instance, you don't have to pay for the transport of electricity, because you have already paid for the hydrogen transport costs. Nor do you have to strengthen the electrical connection in your home for heating. You can also make flexible use of the fuel cell, and produce electricity for the neighborhood whenever there is a shortfall. And the fuel cells should even be able to produce electricity to charge the batteries of electric cars, if no other electricity is available.

If we look further into the future, then you might expect to have reversible fuel cells that can work as a fuel cell and electrolyzer. Prototypes of such cells are already available [9]. If there is a surplus of solar power, you can use a reversible fuel cell to produce hydrogen and feed it into the hydrogen network. In this way you can absorb peaks in electricity production, without overloading the electric grid or dis-connecting the solar panels when they supply too much electricity on a sunny day.

3

CLEAN ENERGY, MATERIALS AND FOOD FROM THE DESERTS

Deserts are not only blessed with a large supply of sun and wind, they also offer plenty of space. This allows you to produce cheap green hydrogen in the deserts. If you do this by using pure water from seawater, then you can also produce drinking water and irrigation water. And from the residual product (brine), you can recover all sorts of chemical products and elements. With fresh-water, sunlight, and fertilizers, you can even develop agriculture in the deserts. The sky is the limit!

Solar and wind energy from the deserts

Deserts have not only the highest solar irradiation in the world, but the wind also blows especially strongly in many deserts. Deserts furthermore occupy extensive surfaces of land where not many people live. They are therefore potentially the suppliers of sustainable energy for the world, in the form of green electricity and hydrogen.

Inexhaustible sources of sun and wind

Desert areas, like those in the Middle East (Saudi-Arabia, the United Arab Emirates, Kuwait, Qatar, Oman, Iran, and Iraq) and North Africa (Libya, Algeria, and others) currently supply the world with oil and gas, and still hold enormous oil and gas reserves. If you look at the huge and inexhaustible sources of solar and wind energy in these areas, the transition from oil and gas to renewable energy is however self-evident [37]. The world's total primary energy use in 2019 amounted to 168,000 TWh, or 606 EJ [30]. You could produce the same amount of energy with solar panels installed on less than 10% of the Sahara's surface. In addition, the wind speeds at many places in the deserts are high, so that you can also generate cheap wind energy there.

You can produce enough solar energy for the whole world on about 10% of the Sahara's surface

In the future, the deserts can also therefore comfortably provide the world with cheap and reliable sustainable energy. The costs of solar electricity, and probably also of wind electricity, could drop in the foreseeable future to less than 1 dollar cent per kWh [45], [46], [47]. The challenge is not only to transport this cheap sustainable energy to local communities, villages, and cities, but also to export it to countries farther away.

You can of course also use electricity from the desert in the region itself. But by converting electricity into hydrogen, you can also store this solar and wind energy cost-effectively and on a large scale, and transport it over large distances. You produce hydrogen through the electrolysis of water, making use of solar- and wind-generated electricity. But how do you get water in the desert? And what is the use of hydrogen to the local communities, towns, and cities?

Stimulus for local communities

If we tackle the development of sustainable energy in deserts smartly, it works both ways: on the one hand, the production of green electricity and hydrogen is used for the benefit of the local communities, towns, and cities, and, on the other, it is exported in the form of hydrogen or derived hydrogen products to other parts of the world. Green electricity and hydrogen in desert areas are in fact not only cheap, clean energy carriers for industry, transport, power, lighting, and heating/cooling. They can also provide for a good, reliable water supply and enable the production of a range of agricultural products, biochemical products, fuels, and materials.

Deserts as source for clean energy, materials, and food

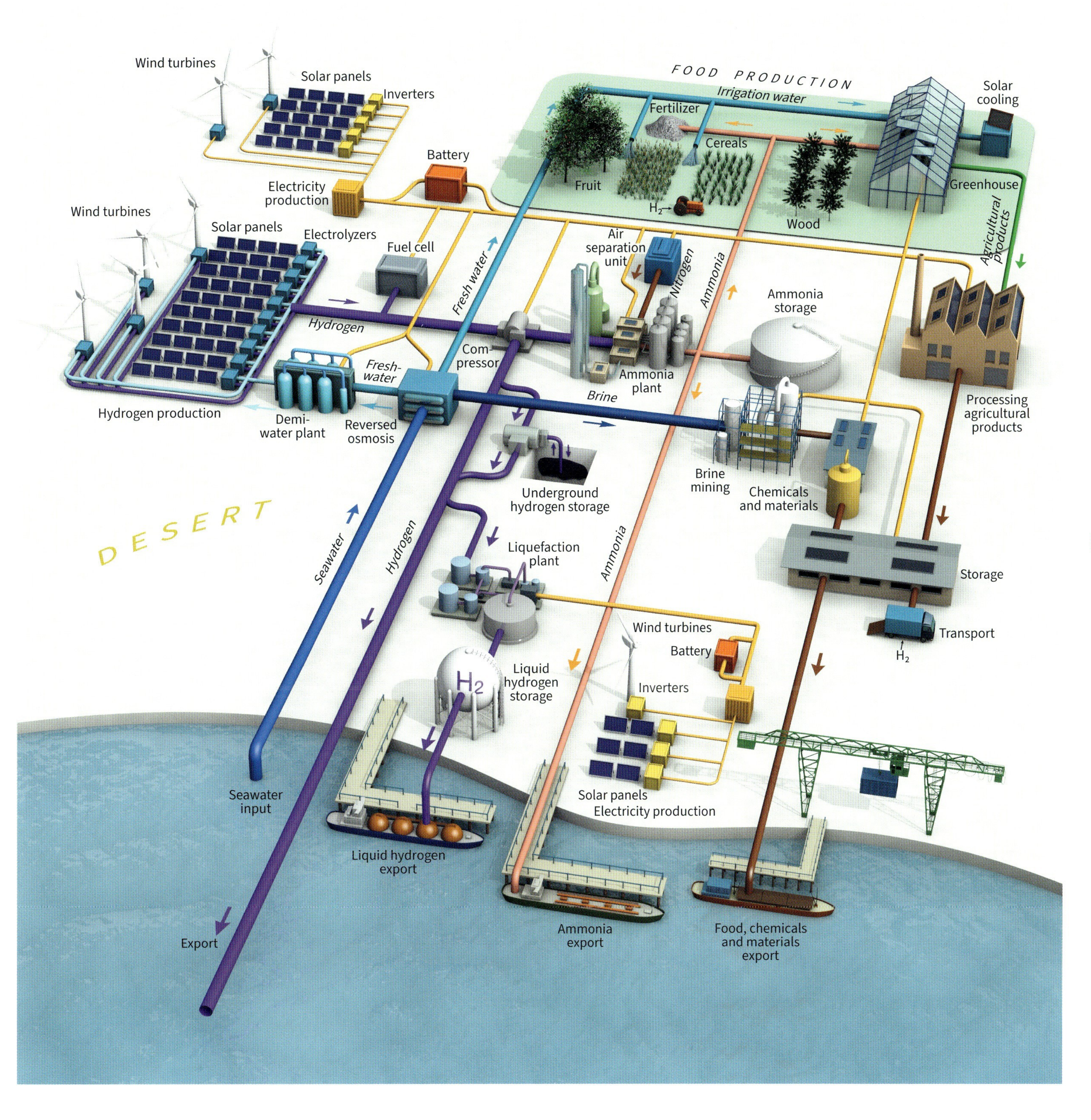

Solar electricity production

The production of solar electricity has grown enormously over the last years, in part because of the installation of large-scale solar farms in desert areas. The largest farms are tens of square kilometers in size and have a capacity in the order of several gigawatts.

Solar panels produce electricity through a process in which the incoming solar irradiation knocks electrons loose in a material, usually silicon. These solar panels supply direct current and are connected in series in order to achieve a higher voltage. An inverter in the solar farm converts the direct current from a number of rows, or "strings", of solar panels, into alternating current at an even higher voltage. The alternating current from the inverters in a solar farm is then transported by infield cables to a central transformer station, where the voltage is increased to a level where the solar power can be transported over larger distances, without significant losses, to cities and industry. The operational time, expressed in full-load hours per year (a unit for the effective annual output), for solar farms in desert areas is between 2,000 and 2,300 hours—more than twice as much as in Germany. A solar farm of 1 GW thus produces an annual electricity output between 2 and 2.3 TWh.

The Mohammed Bin Rashid Al Maktoum Solar Park in Dubai, with 3 GW of solar panels and CSP solar energy system [72]

Wind farm at Golmud, Qinghai province, China [73]

Besides solar panels there are also solar power plants that collect incoming solar rays with mirrors. These concentrating solar power (CSP) plants concentrate the collected sunlight in a pipeline or a central receiver, where it heats a fluid to high temperatures. As in a traditional electric power plant, electricity is then produced through a thermal cycle. The production of electricity with these systems is more expensive than with solar panels. But their advantage is that you can store the hot liquid in a vessel, and use it to produce electricity at night as well.
It is expected, however, that batteries will become so cheap that they can be deployed to store electricity generated by solar panels, so that it can be used at night. The question is whether the use of batteries in the deserts will become cheaper than CSP in solving the day/night storage problem.

A solar panel in the Sahara produces 2 to 2.5 times more electricity than a solar panel in Germany

Wind electricity production

The wind blows hard in several desert areas, both in the deserts themselves and in desert areas on the coast, in the form of powerful sea breezes. The latter occur in California and around the Red Sea, for example. Sea breezes can produce high wind speeds when warm air rises above the desert, creating a low-pressure area. Cold air above the sea then flows to this low-pressure area. The highest wind speeds occur in the afternoon. Since one can use this to produce electricity primarily in the afternoon, evening, and at night, wind and solar energy complement each other well in time: during the day, the most solar energy, and in the afternoon, evening, and at night, the most wind energy.

In recent years wind turbines have become increasingly large. Today, capacities up to 15 MW are indeed possible. Because of the wake effects (a wake of swirling air is created behind the wind turbines), large wind turbines will be set up more than 1 km apart. A wind farm of 1 GW or more will soon take up tens of square kilometers. But, in contrast to solar farms, there is still plenty of space between the turbines that can be used for other purposes, such as agriculture.

The electrical system configuration of a wind farm is comparable to that of a solar farm. Wind turbines deliver "irregular" alternating current, the frequency of which fluctuates with the rotation speed. An inverter in the wind turbines converts the alternating current into direct current with a low voltage, and then into a "uniform" alternating current of 50 or 60 Hz with a higher voltage. A central transformer station raises the voltage further, making it possible to transport the electricity over larger distances with lower loss. At a good site, the operational time of a wind farm ranges from 4,500 to 5,500 full-load hours. A wind farm of 1 GW thus produces between 4.5 and 5.5 TWh of electricity annually.

Solar and wind hydrogen production

You can also convert solar and wind energy into hydrogen, enabling its large-scale storage and transport over long distances in a cost-effective manner. The prevalent idea today is to set up a single, large-scale electrolyzer installation next to a large solar and/or wind farm with which to produce hydrogen. But the cost-effectiveness can be improved by directly using the direct current electricity produced by solar panels and wind turbines in electrolyzers to produce the hydrogen. In this way, we "integrate" the electrolyzer with each wind turbine or string of solar panels.

The costs of hydrogen production can be reduced by integrating a string of solar panels or a wind turbine with the electrolyzer

Electrolyzer installations are built from modular units: stacks with a capacity between 1 and 10 MW, which work on direct current. Solar panels produce direct current and, following the first conversion step, wind turbines do as well. But if you now integrate the electrolyzer stacks, instead of the DC/AC inverters, in a solar farm or a wind turbine, then you avoid several electricity conversion

Hydrogen production farm, modularly built with strings of solar panels with integrated electrolyzer stacks

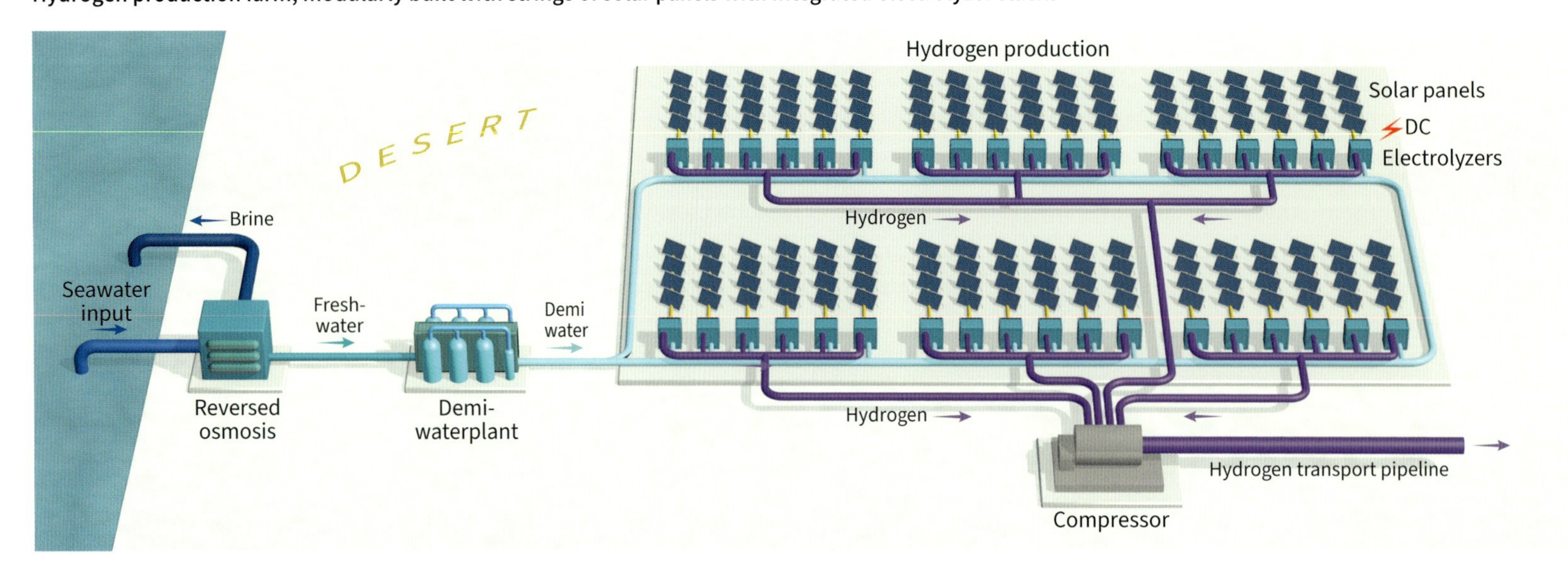

steps. Furthermore, you can then utilize standard solar- and wind-hydrogen products, and be able to develop solar- and wind-hydrogen farms. The costs of hydrogen production will be considerably reduced thanks to this integrated approach.

The production of hydrogen requires water. This means that you need to install an infield pipeline network to transport demineralized water to the electrolyzers. In addition, an infield network of hydrogen pipelines is required to transport the hydrogen from the electrolyzers. On the other hand, no infield electric grid is needed, though a small power source—in the form of a battery, for instance—is required at the electrolyzer stacks. What still has to be done centrally is the gas cleaning, compression (to feed hydrogen with the right specifications and at the right pressure into the hydrogen transport network), and the production of demineralized water.

TECHNOLOGY FOR THE FUTURE

Photolysis

Direct production of hydrogen from sunlight and water

A new development is the direct use of sunlight, photons, to split a water molecule into hydrogen and oxygen. This means that you skip an intermediary step, that is, electricity production. The process is called photolysis, or photo-electrochemical conversion. Several universities in the world are investigating and improving this process, and it appears that increasingly higher efficiencies are possible [74]. This development is similar to the rapid improvements in photo-voltaic solar panels in the 1980s and 1990s. Today, the efficiency of photolysis is already comparable to solar panels combined with electrolyzers. But there are still problems to be overcome. For instance, the panels use rare and expensive catalyst materials, and the efficiency of the process declines over time. A few start-ups have already put these photo-electrochemical panels on the market.

SOLHYD solar panel that produces hydrogen directly from sunlight [75]

Water in the deserts

The production of hydrogen requires clean, demineralized water. But since it is usually not available in desert areas, we can bring in seawater and use it to make demineralized water onsite. The costs involved amount to only a small percentage of the total hydrogen production costs.

Inexhaustible sources of solar and wind

Many countries with large desert areas are affected by water scarcity. It rains little in deserts and, when it does, the rains are often concentrated in specific periods. Some desert areas have subsurface water reservoirs in the form of aquifers, but this is far from the case everywhere. In any event, it is best to leave those that do exist as undisturbed as possible. Water can be transported by pipeline over great distances. Indeed, transport over hundreds up to thousands of kilometers is technically quite feasible and relatively inexpensive: 100 km of water transport costs roughly € 0.05 per cubic meter (= 1,000 liters) [76].
The sea is often relatively close to desert areas, and can serve as a source of water. And like freshwater, you can transport seawater over large distances cheaply. Through the application of reverse osmosis one can make freshwater from this seawater. This procedure is already used to produce drinking water in many countries affected by water scarcity.

Making freshwater through reverse osmosis

Reverse osmosis can be used to make freshwater from seawater. The treatment separates salts and other impurities from seawater or wastewater. This is done using pressure and a membrane, through which water molecules are pressed. Apart from the freshwater, the process also generates a saline residual stream: the brine.

As the name indicates, reverse osmosis is the reverse form of osmosis. Osmosis is the natural tendency of water containing dissolved salts to flow through a membrane from a lower to a higher salt concentration. This process occurs in many places in nature. Plants use osmosis to ingest water and nutrients from the soil. The kidneys of humans and other animals use osmosis to remove water from the blood. And one can also use osmosis to generate electricity, by having freshwater flow through a membrane into saline water.

To use reverse osmosis to make drinking water from seawater, you need about 2.50-3.50 kWh/m^3 of electricity. The total costs—for energy, investment, and maintenance—of producing 1 m^3 of drinking water with this method are € 1.00-1.50 /m^3 [77], [78].

Demineralized water

The production of hydrogen via electrolysis requires very clean water: demineralized water. Ordinary drinking water is not clean enough for this purpose. In order to make demineralized water, following the reverse osmosis the water has to undergo a further process, namely, continuous electro de-ionization. This means that the energy use and costs are higher than those for the production of drinking water. To make demineralized water from seawater requires 3-4 kWh/m^3. The cost of producing

Interior desalination installation

The costs of demineralized water for hydrogen production in the desert, using seawater transported over 1,000 km, only represent a few percent of the total hydrogen production costs

demineralized water from seawater is about € 2.00/m³. The transport of 1 m³ of seawater over a distance of 1,000 km costs € 0.50/m³. To produce 1 m³ of demineralized water, you need roughly 2 m³ of seawater. The volume of the residual stream, the brine, is also 1 m³, and will be returned to the sea. All in all, the cost of transporting seawater and brine over 1,000 km is € 1.50 per m³ of demineralized water. When you add € 0.50/m³ for pump energy, then the total cost for the production and delivery of demineralized water to the electrolyzer amounts to about € 4.00/m³ (€ 2.00/m³ production costs, plus € 2.00/m³ transport costs).

The question is how these costs affect the production costs of hydrogen. We can calculate this as well. For the production of 1 kg hydrogen, 9 liters demineralized water are needed in theory. Assuming a loss of 1 liter, then 10 liters of demineralized water is needed for every kg of hydrogen you produce. From 1 m³ (or 1,000 liters) demineralized water you can therefore make 100 kg hydrogen. The costs for demineralized water therefore amount to € 0.04/kg H_2. At a production price of € 1.00/kg hydrogen, the share of demineralized water (produced from seawater and transported over 1,000 km), is only 4%.

Specification of costs of demineralized water from a seawater source, for the production of hydrogen via electrolysis

	Distance from sea (km)	Costs (€/m³ demineralized water)
Transport costs seawater (2 m³ seawater per m³ demineralized water)	1,000	1.0
Production costs demineralized water		2.0
Transport costs brine (1 m³ brine per m³ demineralized water)	1,000	0.5
Other costs (pump energy)		0.5
Total (€/m³ demineralized water)		**4.0**
Costs of demineralized water for production of 1 kg hydrogen (€/kg hydrogen)		**0.04**

Chemical products and materials from seawater

Seawater consists primarily of water, but it also contains many other useful elements, such as potassium for the operation of electrolyzers and batteries. Other elements found in seawater include silicon (as raw material for solar cells), lithium (as raw material for modern batteries), magnesium (as storage medium for hydrogen), and uranium (as source of nuclear energy).

This figure shows the separate elements, such as sodium (Na), potassium (K), and magnesium (Mg), but these are mainly present in the form of salts, such as sodium chloride, or kitchen salt (NaCl), magnesium chloride (MgCl), and potassium chloride (KCl). Centuries ago, these salts were produced by having seawater flow into large salt ponds, and letting it evaporate in the sun. Today, one can use electrochemical processes to make the separate elements (Na, K), but also sodium hydroxide (NaOH) and potassium hydroxide (KOH) from these salts.

Electrolyzer for chlorine production from salt

The production of chlorine from sodium chloride is done through electrolysis. Chlorine factories are therefore electrolyzers that separate salt (NaCl) that is dissolved in water (H_2O) into chlorine (Cl_2) and sodium hydroxide

Composition of 1 kilo of seawater [79]

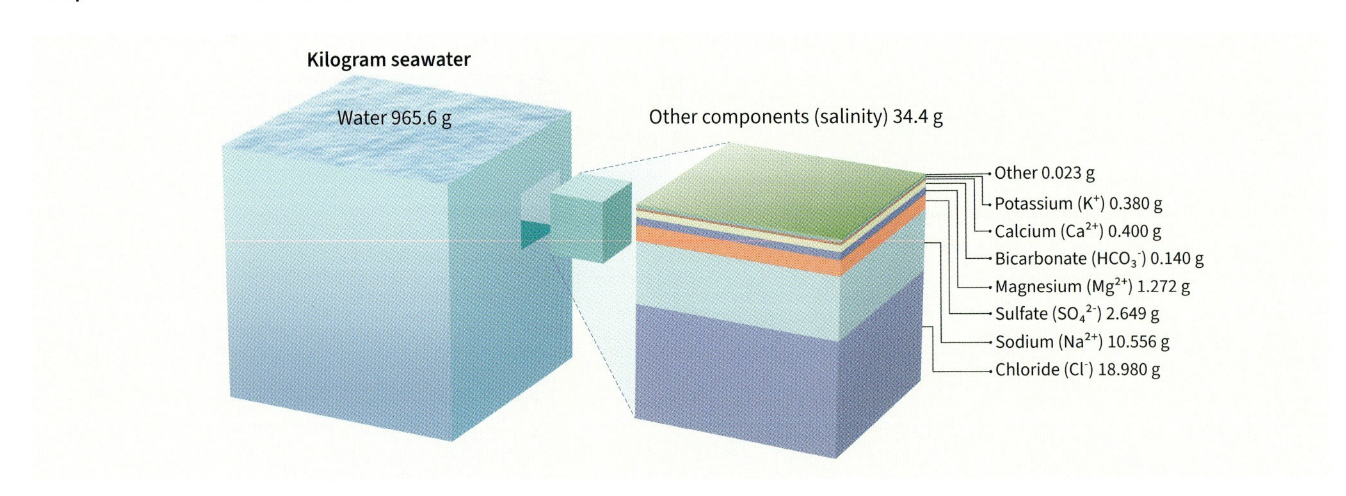

An electrolyzer for the production of hydrogen from water is essentially the same as those used for the production of chlorine from salt

(NaOH). At the same time, water (H_2O) is separated into hydrogen (H_2) and oxygen (O_2)—the by-products of chlorine production. The first electrolyzers were set up more than a hundred years ago, as chlorine factories. An electrolyzer for the production of hydrogen is essentially the same as those used for the production of chlorine. The difference is that to produce hydrogen one uses water that does not contain dissolved salts.

The most commonly used type of electrolyzer for the production of chlorine and hydrogen is the alkaline electrolyzer. This kind of electrolyzer uses electrolytes containing water and potassium hydroxide. Alkaline batteries also use this electrolyte, which occasionally needs to be refreshed. This is why it is interesting that one can use seawater to produce potassium hydroxide, a chemical substance that is essential for both alkaline electrolyzers as well as alkaline batteries.

Valuable elements in seawater

A small fraction of seawater consists of elements other than water. Many of these elements have useful applications. For example, the following elements are of key importance for energy technologies: silicon (raw material for solar cells), lithium (raw material for modern batteries), boron (candidate for hydrogen storage), rubidium (raw material for ultra-thin batteries), and uranium (source for nuclear energy) [80].

The concentrations of these elements are in the order of micrograms to milligrams per liter of seawater. The concentration of lithium in seawater is for instance roughly 0.1 mg per liter. For 1 kg of lithium one therefore needs 1 million liters of seawater. And because lithium is not present in seawater in free form, a chemical process has to be applied to extract lithium from seawater.

You can use seawater to produce potassium hydroxide, an essential substance for the operation of alkaline electrolyzers and batteries

Brine mining

The brine that remains after reverse osmosis (half a liter per liter of seawater), can be concentrated more by dewatering it even further, through evaporation for example. Through a sequence of processes and techniques, you can then recover a variety of salts, chemical substances, and valuable elements from the concentrated brine. What is ultimately left is a water stream with high concentrations of elements that have yet to be extracted. This brine mining process has been the object of numerous studies. These show that the costs are relatively high in relation to the returns. A positive business case may possibly be developed if larger water streams are treated, or if the prices of the substances to be extracted increase further [81].

Zero liquid discharge

Many reverse osmosis plants currently discharge the brine through a pipeline into the sea, where it quickly mixes with seawater. If large volumes are discharged, or if the discharge is done in fragile nature areas, such as coral reefs, it can have a damaging impact, including coral bleaching. For this reason zero liquid discharge—absolutely no discharge of the brine into the sea—is being worked on. This is for instance the underlying principle followed in building NEOM, a new city in Saudi-Arabia [82].

To achieve this objective, all of the water has to be removed from the brine. This can be done for instance by evaporating the brine in the sun, which is not a problem in the desert. The remaining solid material can then be stored, with a view to the extraction of valuable elements from it at a later stage. If you discharge no brine into the sea, then no investments need to be made in a return pipe, pumps, and a discharge station.

Agriculture in the deserts

To grow food you need sunlight, water, and nutrients, such as ammonia. Even in a desert, you can have these in abundance. You can power agricultural machinery with hydrogen or ammonia, and you can cool greenhouses using solar heat. A stimulus for both the local economy and for enhancing sustainability in desert areas.

Everything you need to grow food you can make available in desert areas. To grow trees and crops, you need sunlight, water, and nutrients. There is plenty of sunlight in deserts, but not of water or nutrients. If you enlarge the pipeline for the transport of seawater for the electrolyzers a little, then you can make not only demineralized water, but also drinking water and irrigation water. The only thing that would still be missing are the nutrients (nitrogen, phosphorus and potassium), for instance, in the form of fertilizers. The primary ingredient of these is however ammonia, and that you can make from hydrogen and nitrogen from the air.

Production of ammonia in deserts

Ammonia is produced through the Haber-Bosch process. Under high temperatures and pressures, the process makes ammonia (NH_3) from hydrogen (H_2) and nitrogen (N_2). Some of the hydrogen produced in desert areas can be used for this purpose. The nitrogen needed can be drawn from the air, which after all is 80% nitrogen. This is done using an air separation unit that runs on cheap solar and wind electricity. Ammonia can therefore be produced in the desert not only at a low cost, but also without CO_2 emissions.

Ammonia as a fuel, for energy storage, and for export

You can use ammonia to make fertilizer, but ammonia is also an energy carrier as well as a chemical product. Because liquid ammonia has a high energy density, and it already liquefies at a temperature of -33 °C or a pressure of 8 bar, ammonia is very suited for the storage of energy in tanks. Ammonia is also a useable fuel for diesel engines and gas turbines. Its combustion releases no CO_2, but it does emit nitrogen oxides (NO_x); these can however be captured with a catalytic converter.

You can also effectively and cheaply transport liquid ammonia by ship or pipelines [83]. Ship transport is already practiced all over the world. This means that one can ship green hydrogen in the form of green ammonia worldwide. Pipeline transport is also already being done. There is an ammonia pipeline network in the United States that runs from the south (New Orleans), through the Midwest, up to the north, over a distance of about 3,200 km. Because this pipeline runs through the country's corn belt, the farmers can, so to speak, tap into the ammonia directly [84].

Agricultural products in the desert

We have shown that, besides sunlight, you can also have water and nutrients available in the desert. But which agricultural crops can you grow with these? The surprising answer is: practically all of them. On the desert sands one can grow trees for wood production, but also fruit trees, olive trees, and grains. And vegetables, flowers, and plants in greenhouses, which, because of the extreme heat, you need to cool. It would also be interesting to plant fast-growing crops like elephant grass, which can be used as a feedstock for a number of bio-chemical products and synthetic fuels.

Working the land and processing agricultural products

To practice agriculture, you need to be able to work the land and harvest, process, store, and finally transport the crops. For this, you require agricultural machinery, like tractors, combines, and harvesters. You will then process the agricultural products: make lumber from the trees, flour from the grain, and juice from the oranges. These end-products have to be stored (fruit and vegetables in coolers), and finally transported to their destination, for example, in trucks. You can power all this mechanical activity using solar and wind energy. Mobile agricultural machinery, like tractors, combines, and trucks, can run on hydrogen—or, in an initial phase, on ammonia. Stationary processing machines, for example, for sawing, milling, and sorting, run on electricity. And cooling can be done using solar heat, driven by an absorption cooling machine.

Desert oasis

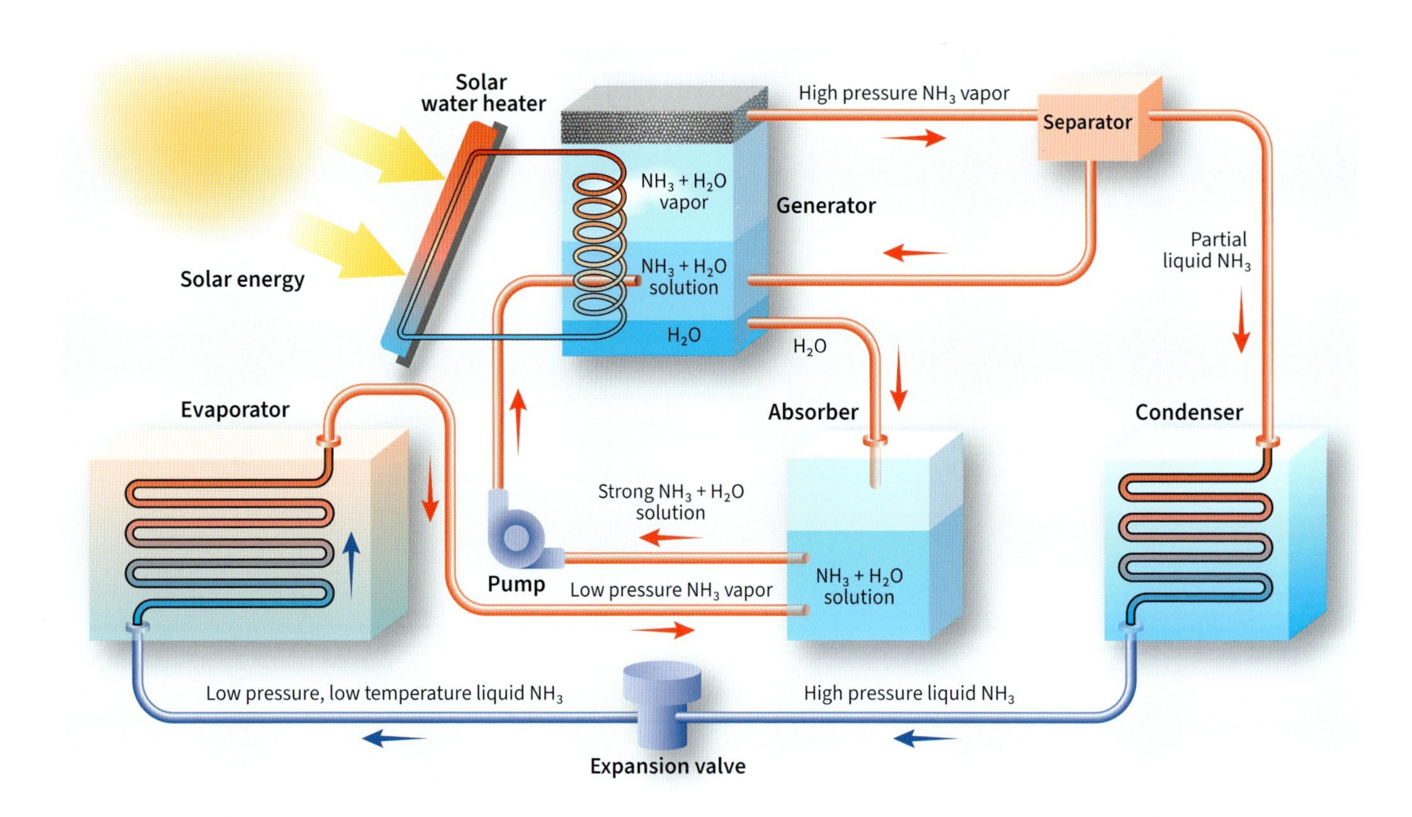

A solar-heat powered cooling machine [85]

Solar energy for absorption cooling

It may sound somewhat contradictory, but in the desert one can cool greenhouses, cold stores, halls, and large buildings with solar-heat absorption cooling. Absorption cooling systems come in various forms, but evaporation absorption machines are of most interest for desert applications, since they make use of solar heat and require only a single pump. An evaporation absorption machine consists of four basic components: an evaporator powered by solar energy, an absorber, a generator, and a condenser. A refrigerant flows from the condenser to the evaporator, via the absorber to the generator and then back again to the condenser. The cooling of the cold store or the building takes place at the evaporator. The evaporation requires energy, and this is extracted from the cold store or the building. This form of cooling is comparable to that of a refrigerator, except that a refrigerator uses a different refrigerant and electricity instead of solar heat. A very common refrigerant is water-diluted ammonia. And ammonia is actually one of those products that we already produce on a large scale in the desert from green hydrogen.

Stimulus for local economy and export

The availability of cheap sustainable energy (in the form of green electricity and green hydrogen), and of sufficient reliable freshwater, can provide a major stimulus to a local sustainable economy. And since there is also plenty of space available, you can also start producing for export: not only sustainable energy (in the form of hydrogen), but also ammonia, and synthetic fuels like aviation fuel or methanol. The export of sustainable agricultural products, green chemical products, and materials originating in brine, are also possibilities.

If you enlarge the seawater pipeline, then you can make both demineralized water for hydrogen/fertilizer production and irrigation water for agriculture

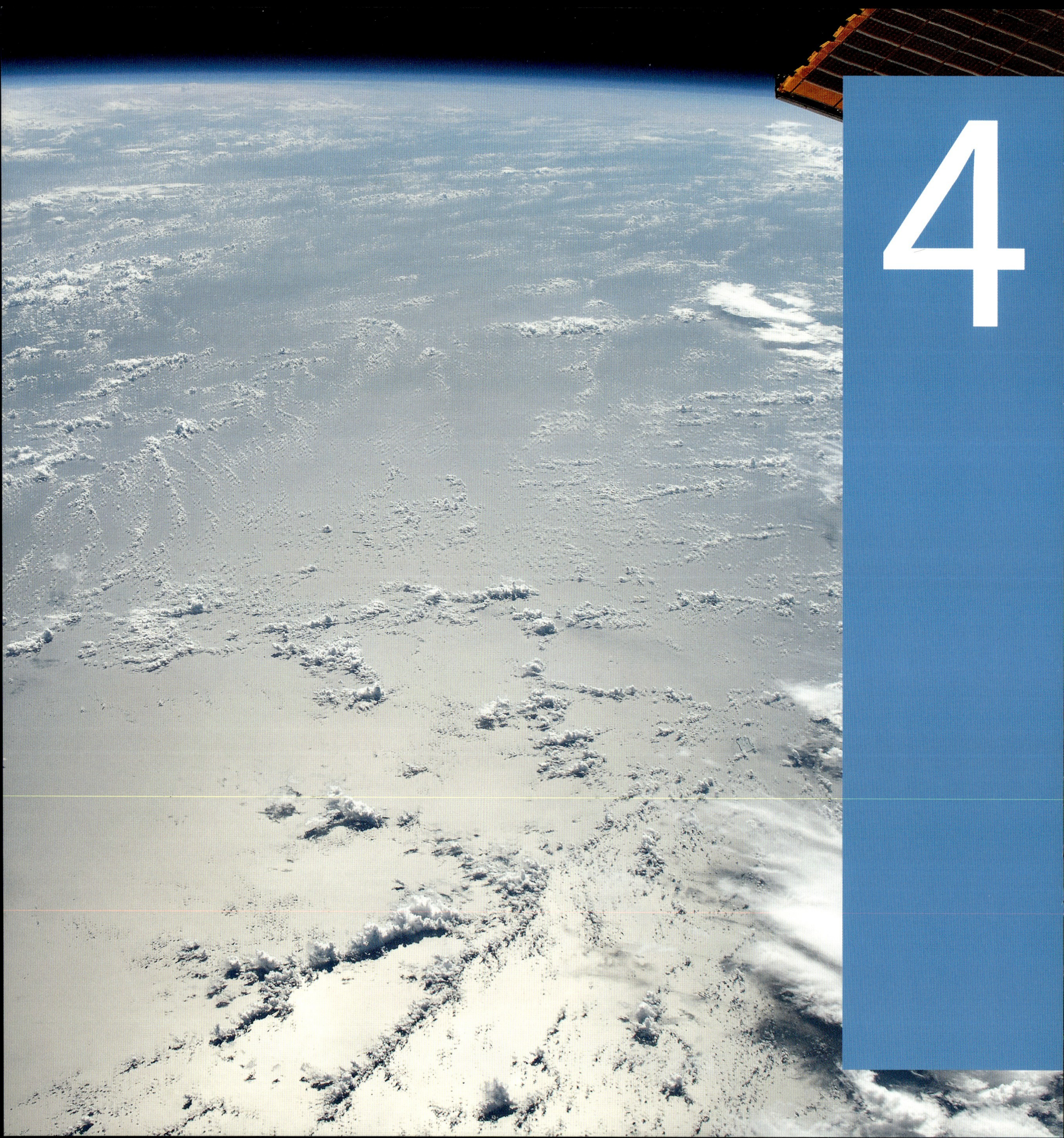
4

CLEAN ENERGY, MATERIALS AND FOOD FROM THE OCEANS

Oceans cover 70% of the surface of the Earth. Above and inside this gigantic water reservoir there are more good sustainable energy sources than on land. Not only are the oceans the source of water, they are also a source of life, food, and energy. The oceans can therefore, just like the deserts, potentially provide for our needs in sustainable energy, drinking water, food, chemical products, and materials.

Large sustainable energy potential

The oceans not only cover a gigantic surface, they are also areas where you can generate huge amounts of sustainable energy. In addition, they offer great potential for the production of synthetic fuels, chemical products, materials, water, and food.

Toward a clean, affordable energy production

Not only on land, but also at sea, all over the world large volumes of crude oil are already being extracted. The top-10 offshore oil and gas producing countries in 2017 were: Saudi-Arabia, Norway, Qatar, Iran, Brazil, the United States, Mexico, the United Arab Emirates, the United Kingdom, and Angola [86]. Much of these extraction activities are now conducted in relatively shallow waters. Considerable oil and gas reserves remain in deeper waters, but the deeper the reservoir, the more difficult and expensive is the extraction.

You can meet the world's total primary energy use needs with wind turbines on less than 2% of the Pacific Ocean

Sustainable energy sources on the sea are far more accessible. You can generate wind energy directly above the water surface, wave and current energy in the upper water layer, and solar energy using floating panels on top of the water. Biomass, in the form of seaweed, also floats on the surface. To use seawater for cooling and for ocean thermal energy conversion (OTEC), you need to reach hundreds of meters deep into the sea, but not quite to the seabed. For almost all of these forms of sustainable energy you need a considerable sea surface, although this represents only a modest portion of the total ocean surface, which covers more than 70% of the planet. Thus, less than 2% of the surface of the Pacific Ocean would suffice to meet the world's total primary energy use needs (in 2019 this was 168,000 TWh, or 606 EJ [30]), if you install one large floating wind turbine every kilometer [29].

In the future the oceans can easily supply the world with sustainable energy. But the generation of sustainable energy at sea is not as simple as it is on land. And the transport of renewable energy on the oceans to areas where there is an energy demand is even more difficult than in the case of deserts. But the oceans also have some big plus points: there are numerous sustainable energy sources of high potential, there is space in abundance, and human activities are limited.
If we tackle the development of sustainable energy on the oceans smartly, we can produce green electricity and hydrogen there. Furthermore, we can also use the oceans for cooling and for the production of synthetic fuels, chemical products, materials, water, and food. The question is how we can do all this in a clean and affordable manner.

Good moment for the use of energy potential

On land we are capable of producing very cheap solar and wind energy at the right locations. And thanks to the exploration of fossil energy at sea, and even in the deep sea, a great deal of experience and knowledge has been

Oceans as source of sustainable energy, water, food, chemical products, and materials

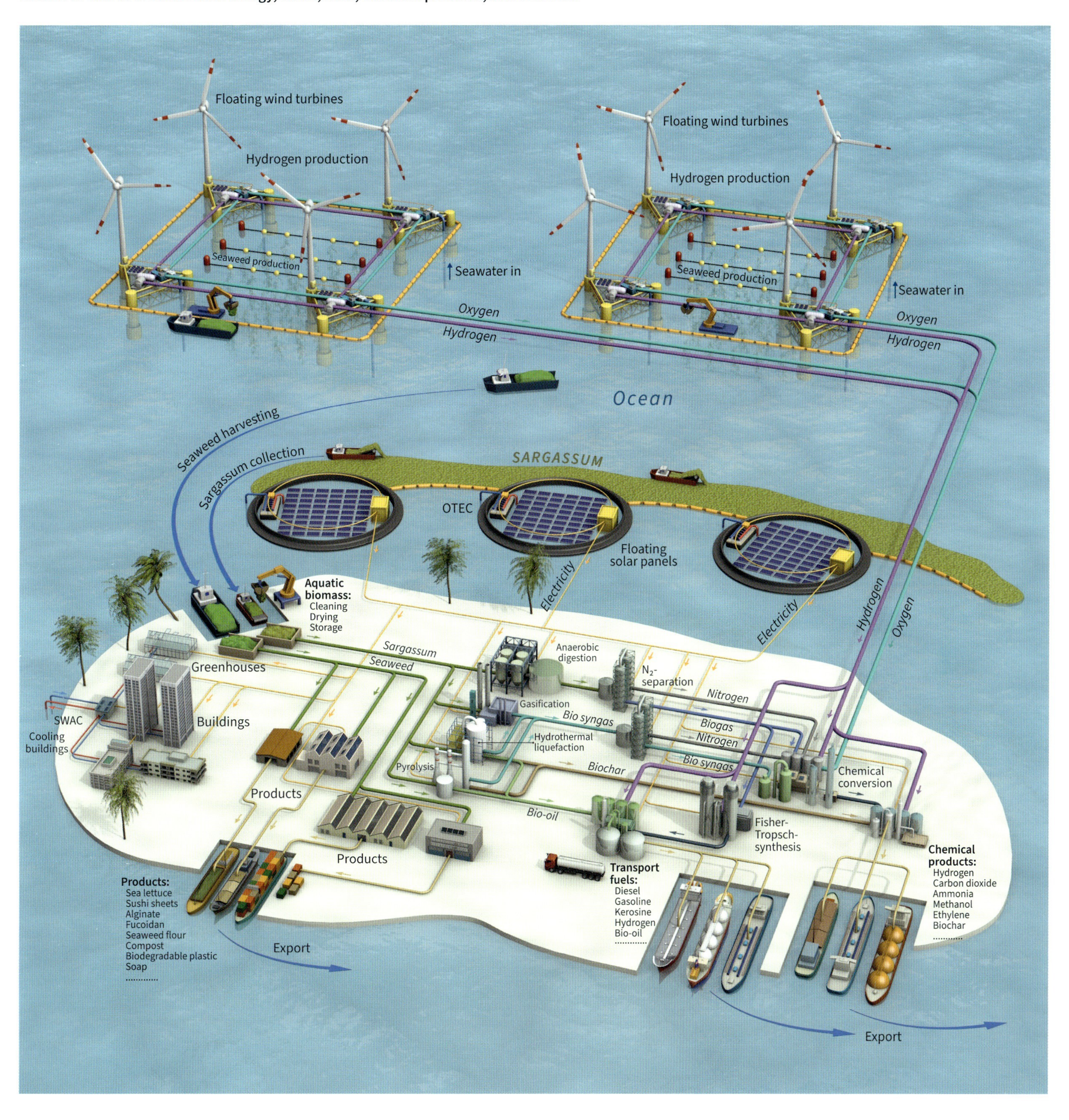

developed on the construction of complex technical installations on oceans and on the seabed. The first steps have also been taken with offshore wind farms close by the coast. Currently we are building the first floating wind farms and developing floating solar farms. Various other sustainable energy technologies are either under development, in the testing phase, or already being realized on a small scale.

With this background in mind, now is an appropriate time to begin exploiting the terrific sustainable energy potential of the oceans. Not by examining each sustainable energy source separately, but by developing different sustainable energy sources in an integrated fashion, accompanied by conversions to usable energy carriers, chemical products, and materials. At the same time, the necessary infrastructure also needs to be developed and storage facilities installed.

Building wind turbines at sea

Thanks to the exploration of fossil energy, there is a great deal of experience and knowledge on the construction of complex technical installations on oceans and on the seabed

Offshore wind energy

If you use floating wind turbines, then there is an enormous potential for offshore wind farms in deep waters. The question is how can we transport the generated energy to land, and how can we best make use of the sea surface between the wind turbines.

It is above the oceans that wind speeds are the highest. The wind there is not hindered by vegetation or buildings, but blows over a flat water surface. When it blows strongly it creates waves. Part of the wind's energy is in this way converted into wave energy.

In 2002, the first large-scale offshore wind farm, Horns Rev 1, was built in Denmark, with a capacity of 160 MW. Each wind turbine in this farm has a capacity of 2 MW, and they are installed in sea sections with a maximum water depth of 14 meters. Twenty years later, the largest offshore wind turbines have a capacity of 15 MW, with rotor diameters of more than 200 meters. What does a wind farm with these turbines look like? And how is the energy brought to land?

Floating wind turbines

Today's offshore wind turbines are installed on a foundation in the seabed. The costs of these foundations increase with the depth. At a water depth of more than 50 meters, it is more attractive to mount the wind turbine on a floating support structure that is moored to the seabed [87]. Installing wind turbines in deep-water areas and farther away from the coast has a number of benefits. The wind there is often more powerful and consistent, so that the turbines can produce more energy [88]. Moreover, the installation costs can be lowered [89].

There are different substructures for floating wind turbines. The key difference concerns the distribution of the weight under water that sustains the turbine. One of the big advantages of the semi-submersible and tension-leg platform types is that they can be entirely assembled in a harbor, and then be towed out to their destination [87]. This significantly reduces the installation costs.

Three different types of floating wind turbines [87]

Wind turbine being built [90]

Offshore wind turbines have grown quickly over the last decades. Various manufacturers supply wind turbines with capacities of up to 15 MW. The rotor diameter of such wind turbines is more than 220 meters, and their total height is 260 meters. The surface covered by the blades is just under 40,000 m^2, almost equivalent to six football fields. A wind turbine like this actually generates up to 75 million kWh of electricity per year. At good locations, the full load hours are more than 5,500 hours which implies a capacity factor of 65% (the number of full-load hours divided by the number of hours in a year [8,760]).

Floating wind turbines for hydrogen production

Floating wind turbines are intended for the generation of wind energy in deeper seas. But because such locations are often situated far from the coast, the question is how can you transport the generated energy easily and cheaply to shore.

We noted earlier that the transport of hydrogen through pipelines is roughly cheaper by a factor of 10 than the transport of electricity with cables. Moreover, the transport capacity of pipelines is roughly greater by a factor of 10 than that of electricity cables; the capacity of a hydrogen transport pipeline is about 20 GW, while that of an electricity transport cable is about 2 GW.

To produce hydrogen at sea, we can set up an electrolyzer in or next to a floating wind turbine, along with an installation that makes demineralized water from seawater. There is certainly enough space: in or next to the wind turbine's tower, which has a diameter of 10 meters, or on a small platform placed next to a floating wind turbine. By converting the alternating current produced by the wind turbine into direct current, you can then feed the electrolyzer directly. And since the solar cells and batteries supply direct current, the entire system of such a hydrogen-producing wind turbine runs on direct current. In fact, you make a new product: a floating wind-hydrogen turbine, which can be fully assembled in a harbor.

In deep-water areas far from the coast the wind is frequently stronger and more constant, so that wind turbines can produce more energy

The output of this wind turbine consists of hydrogen at a pressure of 30-50 bar. A flexible pipeline connected to several wind turbines, transports the hydrogen to a

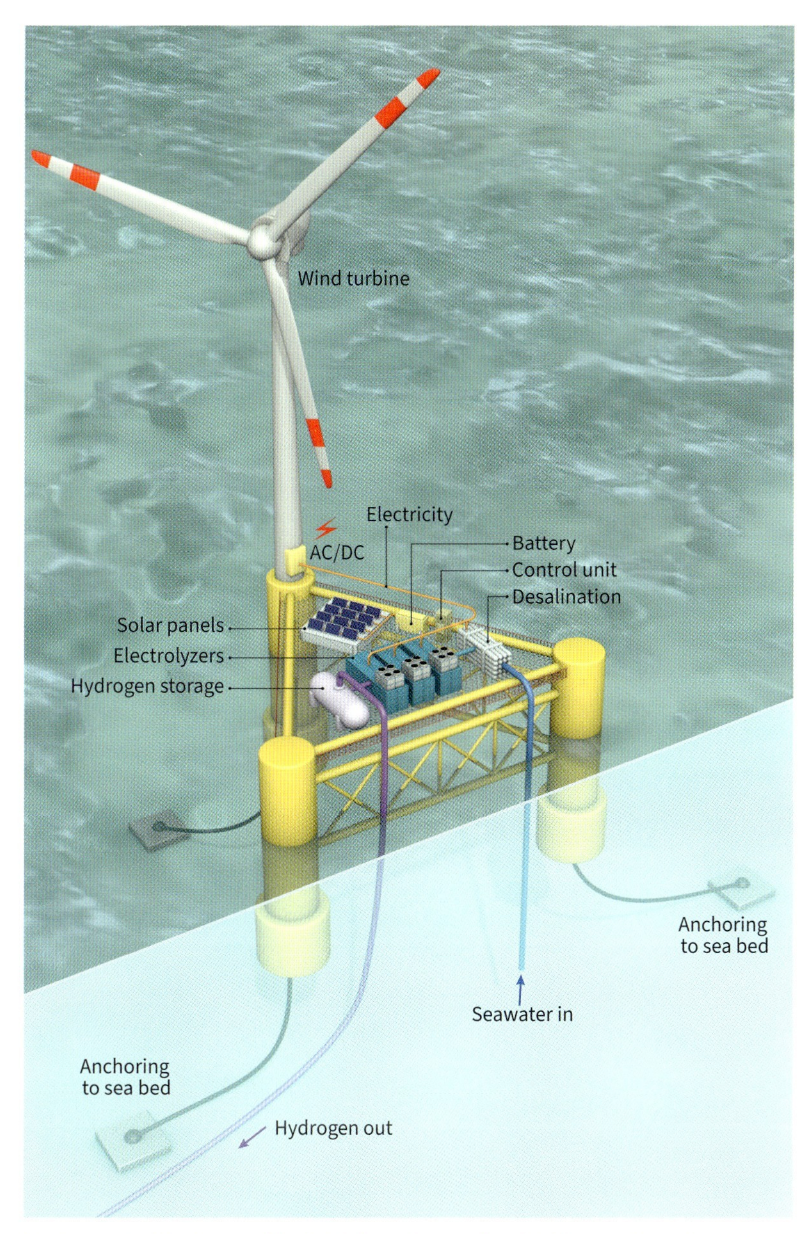

Floating wind turbine with electrolyzer to produce hydrogen (based on ERM design [91]**)**

compressor, which increases the pressure to 80-100 bar, and then pumps the hydrogen to a large pipeline that transports it to shore.
When the wind blows strongly, the wind turbines produce more hydrogen and the compressor requires more electricity. It therefore makes sense to "integrate" the compressor with a (floating) wind turbine, which similarly produces more electricity when the wind blows harder. In addition, a back-up electricity provision is also needed. This can consist of batteries and a fuel cell. The compressor also should be stabilized, which can be effectively done using motion stabilizers [92].

Offshore wind energy produced at more than 100 km from the coast can be delivered to shore more cheaply as hydrogen than as electricity

A floating offshore wind farm of 1 GW with 5,500 full-load hours, linked to a 1 GW electrolyzer capacity, delivers about 110,000 tonnes of hydrogen per year. Less than 10 GW offshore wind-hydrogen capacity thus produces 1 million tonnes of hydrogen per year.

If we now compare the cost price of offshore wind that is brought to land in the form of electricity with cables, with the cost price of offshore wind that is brought to land in the form of hydrogen with pipelines, then, at distances of roughly more than 100 km from the coast, the hydrogen delivered to shore is cheaper than the electricity delivered to shore [13]. True, the hydrogen route requires that you invest in an electrolyzer and also involves a conversion loss (conversion of electricity into hydrogen), but this is compensated by the fact that fewer conversions are required (between alternating and direct current, and to higher and lower voltages), and the hydrogen transport costs are lower than those for electricity.

Pipeline network for hydrogen transport in the North Sea, predominantly built from existing natural gas pipelines [13]

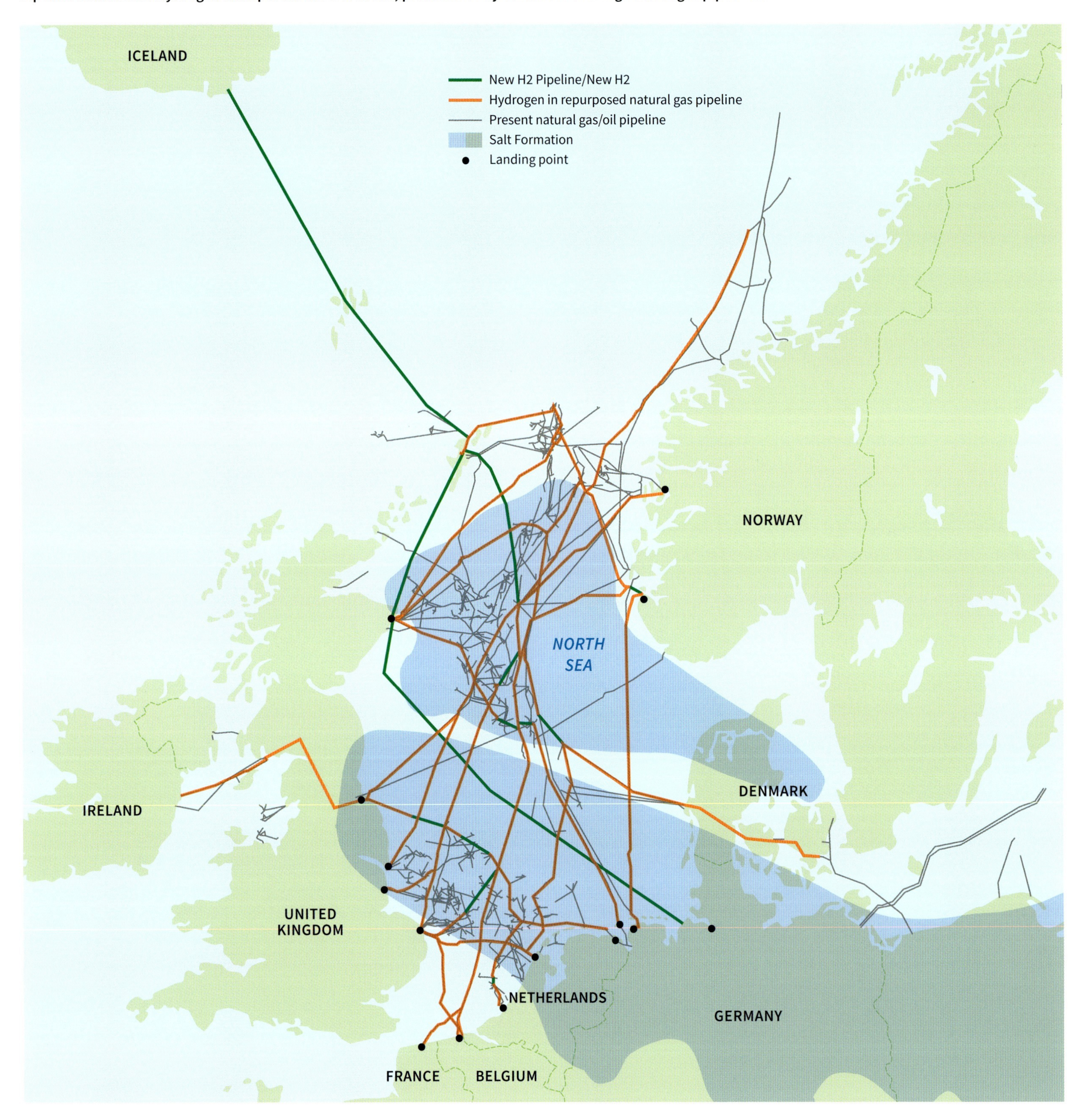

Offshore wind-hydrogen production on the North Sea

Large amounts of oil and gas have been extracted in the North Sea over the last few decades. The gas extraction is still ongoing, but many fields are approaching the end of their production life. What is left behind afterwards is a large natural gas transport network. In recent decades offshore wind farms have also been built in the North Sea, all of them in shallow waters and close to the coast. Currently, attention is being increasingly directed at the development of wind farms farther away from the coast and in deeper waters. In 2021, 8,600 km^2 of the Scottish seabed was auctioned, in the ScotWind seabed tender, for the development of offshore wind farms. In March 2022, 17 offshore wind projects were approved, for a total capacity of almost 25 GW, of which 15 GW in floating offshore wind farms.

Although the 2023 initiatives still only concern electricity production, the possibilities of producing offshore wind energy in the form of hydrogen are under study. The existing network of natural gas pipelines can be employed to transport hydrogen: the pipelines of various North Sea companies are suited for the purpose. On the basis of the existing network and the current developments in offshore wind, particularly floating wind, we have investigated what a pipeline network for hydrogen in the North Sea could look like.

Hydrogen from the North Sea can easily be transported to all parts of Europe through existing natural gas pipelines

We can repurpose more than 15,000 km of existing natural gas pipelines for a hydrogen transport network in the North Sea. A little less than 3,000 km of new hydrogen pipelines would also be needed. The North Sea pipeline network map on page 90 also shows a connection with Iceland. This pipeline can be used to export cheap hydrogen produced from hydro power and geothermal energy in Iceland to Europe. The existing connections to the coast can simply be hooked up to the European Hydrogen Backbone [70]. In this way, hydrogen from the North Sea can easily be transported to all parts of Europe.

At least as important is the possibility of storing hydrogen in the North Sea. The suitability of some empty gas fields for this purpose is being researched. These gas fields are of course already connected to the existing natural gas pipelines. If they turn out to be suited for hydrogen storage, then the existing platforms can also be repurposed to pump hydrogen into and out of the empty fields. Depending on the size, 50,000-500,000 tonnes (2-20 billion kWh [HHV]) of hydrogen can be stored in an empty gas field, primarily for seasonal storage [13], [93].

An alternative is to store hydrogen in salt caverns. Salt formations in many places under the North Sea have the requisite thickness for the construction of salt caverns. You can store 3,000 to 6,000 tonnes (120-240 million kWh) in such salt caverns, which is 10 to 100 times less than in empty gas fields. However, the advantage of salt caverns is that you can load and unload them faster than gas fields. For this reason, you can make good use of them for storage for periods of days or weeks.

It is possible to make hydrogen on the North Sea using offshore wind. Since hydrogen can be seasonally stored in empty gas fields, and daily or weekly stored in salt caverns, a continuous flow (baseload) of hydrogen to shore is possible. Many industries, such as steel and chemicals, require a hydrogen baseload. The hydrogen network we have outlined can transport between 300 and 400 GW of hydrogen to shore. If this is done in baseload, then this amounts to 60 to 80 million tonnes of hydrogen (2,400-3,200 billion kWh [HHV])—about one quarter of the European Union's total energy use [94].

In order to produce this entirely with offshore wind capacity, between 550 and 750 GW of offshore wind capacity is needed. The installation of this wind capability requires between 75,000 and 100,000 km^2 of space, that is,

Kite ships

Flexible energy transport over long distances

With floating wind turbines you can generate wind electricity at deeper water depths, frequently at great distance from the coast. A hydrogen pipeline is needed to bring this energy to shore. But since this sometimes involves distances requiring thousands of kilometers of pipelines, the question arises as to whether other solutions might exist for this energy transport.

One solution is to make use of a kite that can generate electricity at great height, higher than that of a wind turbine, where the wind blows even more strongly. Set up on a ship, the kite can always be flown where the wind is strong. The capacity factor is therefore 80-90%, which is equivalent to 7,000 to 7,800 full-load hours. The electricity generated by the kite is converted into hydrogen and stored under pressure in tanks. When the tanks are full, the kite ship returns to harbor to discharge the hydrogen. The kite ship is flexible and can sail to the harbor where the hydrogen fetches the highest price. This is an advantage compared to wind turbines with pipelines. This technology certainly has potential, but it needs to be developed much more before it can be profitably exploited. A start-up and various technical universities are working on it [95].

Artist impression of kite ships producing hydrogen

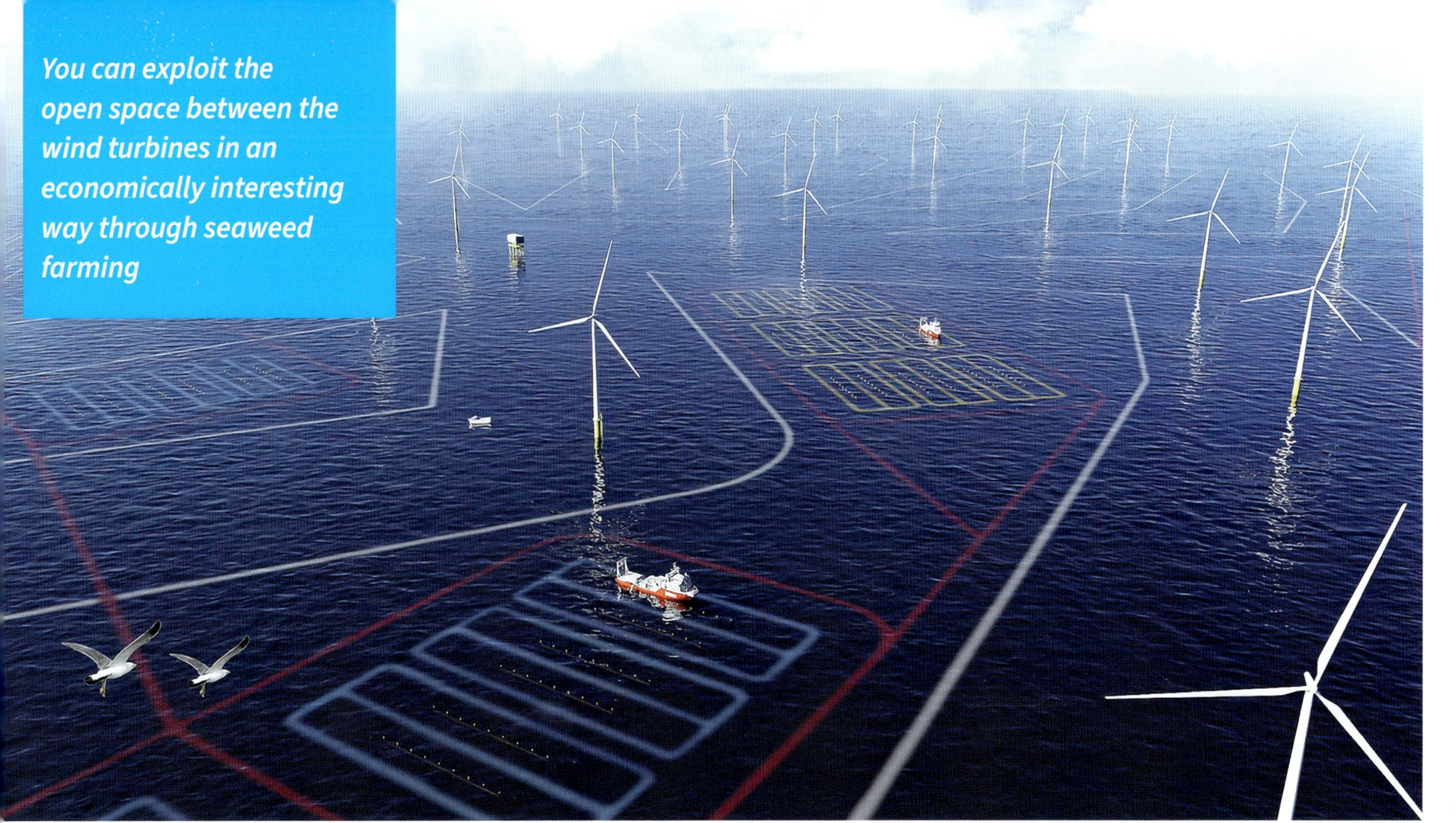

Artist impression of seaweed farming in wind farm [97]

13-17% of the surface of the North Sea. Incidentally, of this area, only 1,500-2,000 km^2 is taken up by the wind turbines themselves. This represents less than 1% of the surface of the North Sea [4].

Cultivating seaweed in offshore wind farms

In wind farms there is lots of space between the wind turbines. Large wind turbines of 15 MW, with rotor diameters of more than 200 meters, can easily be more than a kilometer away from each other. In the space between the turbines you can do other things, such as farm fish, install floating solar farms, or farm crops like seaweed [97]. Seaweed is a food source for humans and animals, as well as a feedstock for chemicals and pharmaceutical products. Seaweed also contains carbon, with which you can, together with hydrogen, produce a variety of synthetic fuels. And seaweed can naturally also be an energy source: you convert it into usable energy carriers.

Seaweed is usually cultivated on long lines that hang in the sea. Wind turbines are ideal constructions for the attachment and mooring of these lines. By farming seaweed in wind farms, you can exploit the open space between the wind turbines in a useful and economically interesting way. Seaweed has additional advantages. It counters acidification and eutrophication, because it absorbs the nitrogen and phosphorus from the oceans. And, as it grows, it also removes carbon dioxide from the atmosphere. In addition, the cultivation areas become breeding grounds for fish.

In an offshore wind farm far from the coast you can produce hydrogen with both wind energy as well as seaweed. If you use seaweed, you can for instance first have it digested. You then apply steam methane reforming or autothermal reforming to it, which allows you to capture 80-90% of the released CO_2. For every

Production of hydrogen and CO_2 by an offshore wind- and seaweed farm

	Assumptions	Production
Wind turbines		
– Full-load hours	5,500 hr/year [98]	
– Efficiency of the electrolyzer	$81\%_{HHV} = 68\%_{LHV}$	
– Installed wind capability	8 MW/km² [4]	
Hydrogen production from wind		**900 tonnes H_2/km²**
Seaweed farming		
– Yield, dry matter	2,000 tonnes/km² [99]	
– Methane production	2,500 kWh/tonnes [100]	
– Efficiency of the steam methane reforming (SMR) with CO_2 capture	$80\%_{HHV}=73\%_{LHV}$ [101]	
– CO_2 capture from SMR (CO_2 from digestion not included)	8 kg CO_2/kg H_2 [101]	
– Surface available in wind farm for seaweed farming	90%	
Hydrogen production from seaweed		**100 tonnes H_2/km²**
CO_2 production from seaweed		**800 tonnes CO_2/km²**

kilo of hydrogen, seaweed thus also produces about 8 kg of CO_2. This does not include the CO_2 from the digestion process. We can now calculate how much hydrogen and CO_2 this wind- and seaweed farm produces per km²; the total amounts to 1,000 tonnes of hydrogen (900 tonnes from wind and 100 tonnes from seaweed), and 800 tonnes of CO_2. Using hydrogen, CO_2, plus nitrogen and oxygen (from the air), we can make all of the synthetic fuels (aviation fuel, diesel, gasoline) and basic chemical products, (methanol, ammonia, benzene, propylene, etc.).
In the sustainable, circular use of the chemical products, the CO_2 is no longer released into the air; in fact, we can even speak of negative CO_2 emissions. Of course, this would also be the case if you stored the CO_2 in an empty gas field or aquifer.

The use of offshore wind-seaweed-hydrogen production farms to produce the equivalent of all the energy for the entire world (168,000 TWh/year) in the form of hydrogen, would require a total ocean surface area of 4.3 million km². This is only 1.2% of the 362 million km² of total ocean surface [102]. In addition, you remove 3,440 million tonnes of CO_2 from the air. If we store all this CO_2 or lock it in products, we can reduce global net CO_2 emissions by about 10% (in 2021 34.9 billion tonnes of CO_2 were emitted [103]). You can therefore regard seaweed farming and processing as a form of direct air capture (DAC): the extraction of CO_2 from the ambient air. And this is probably much easier and cheaper to do than with a technical installation!

Offshore solar energy

Because of the spatial constraints on land, solar panels are also installed on water. An extra benefit: the water cools the panels, whereby they produce a higher output. The question is: can you also make use of solar panels on the sea, in a saline environment, and with high waves?

Floating solar panels on the sea

A number of companies are exploring the potential of floating solar panels on the sea, and solar cell and panel producers have announced plans to collaborate with offshore technology companies [104], [105]. There are also a variety of start-ups and new developments in the field. These are all signals indicating that floating solar panels and solar farms on the sea are becoming a serious option [106], [107].

There are many indications that floating solar farms are becoming a serious option for the production of electricity and hydrogen on the sea

There are many ways of having floating solar panels move with the motion of the water. The first floating systems were derived from land-based systems: solar PV modules were placed on a small floating platform; several floating platforms were then attached to each other. A second design consisted of floats with pontoons. Solar panels were installed on top of these, with the same support structures as used on land. This approach works well on calm waters, such as lakes, but not when the waves are high, as at sea.

The Ocean Sun company came up with an innovative solution, borrowed from fish ponds in fjords and on the sea: solar panels are held together with a hydro-elastic membrane, which, for mooring, is enclosed by a rigid, circular frame [107]. Because the membrane is in contact with the underlying water, the cooling is superior to that of the above-mentioned configurations. In addition, the membrane allows the solar panels to accompany the wave motion, so that the wind resistance is minimized. This configuration can withstand wind speeds of up to 275 km per hour, which is comparable to the wind force of a hurricane.

Compared to land-based systems, the floating solar farms' Balance-of-System (BOS) costs (for the support structure, mooring, and salt protection) and maintenance costs are higher. However, if you generate more energy per surface unit, then you reduce the share of the BOS costs. For floating solar energy systems on the sea, it therefore makes sense to opt for panels with a higher efficiency, even if they are relatively more expensive. The ultimate choice is an outcome of an economic assessment; higher efficiency does not always equal lower costs. But floating solar farms must of course be deployed on the sea wisely, particularly since they limit sunlight penetration into the water, which can impact marine life.

Floating solar panels for hydrogen production

As in the case of wind turbines, we also want to be able to deploy solar farms at a greater distance from the coast. Since hydrogen transport costs are lower than those for electricity, it could be of interest to convert the electricity generated into hydrogen. And because solar panels deliver direct current, you can practically feed it directly into the electrolyzer.

Even if there isn't more solar irradiation on the sea than on land, the output of floating panels can nonetheless be higher by a few percent. This is due to the better (water) cooling of the panels. The number of full-load hours at sea does not differ much from the number of full-load hours on land at the same latitude. At good locations, the number of full-load hours for solar panels is about 2,000. If we connect the solar panels to an electrolyzer with the same capacity, then it, too, has only 2,000 full-load hours. In practice, it makes more economic sense to size the capacity of the electrolyzer at about 70% of that of the solar-cell system, in order to increase the number of full-load hours. In this manner you save considerably on electrolyzer investment costs, while losing only a few percent of the solar panels' output.

Floating solar panels on a hydro-elastic membrane, suited for application with high waves [107]

Even if there isn't more solar irradiation at sea, because of the better (water) cooling the output can nonetheless be a few percent higher there than on land

Floating solar-hydrogen production farm north of Crete that can produce 1 million tonnes of hydrogen [108]

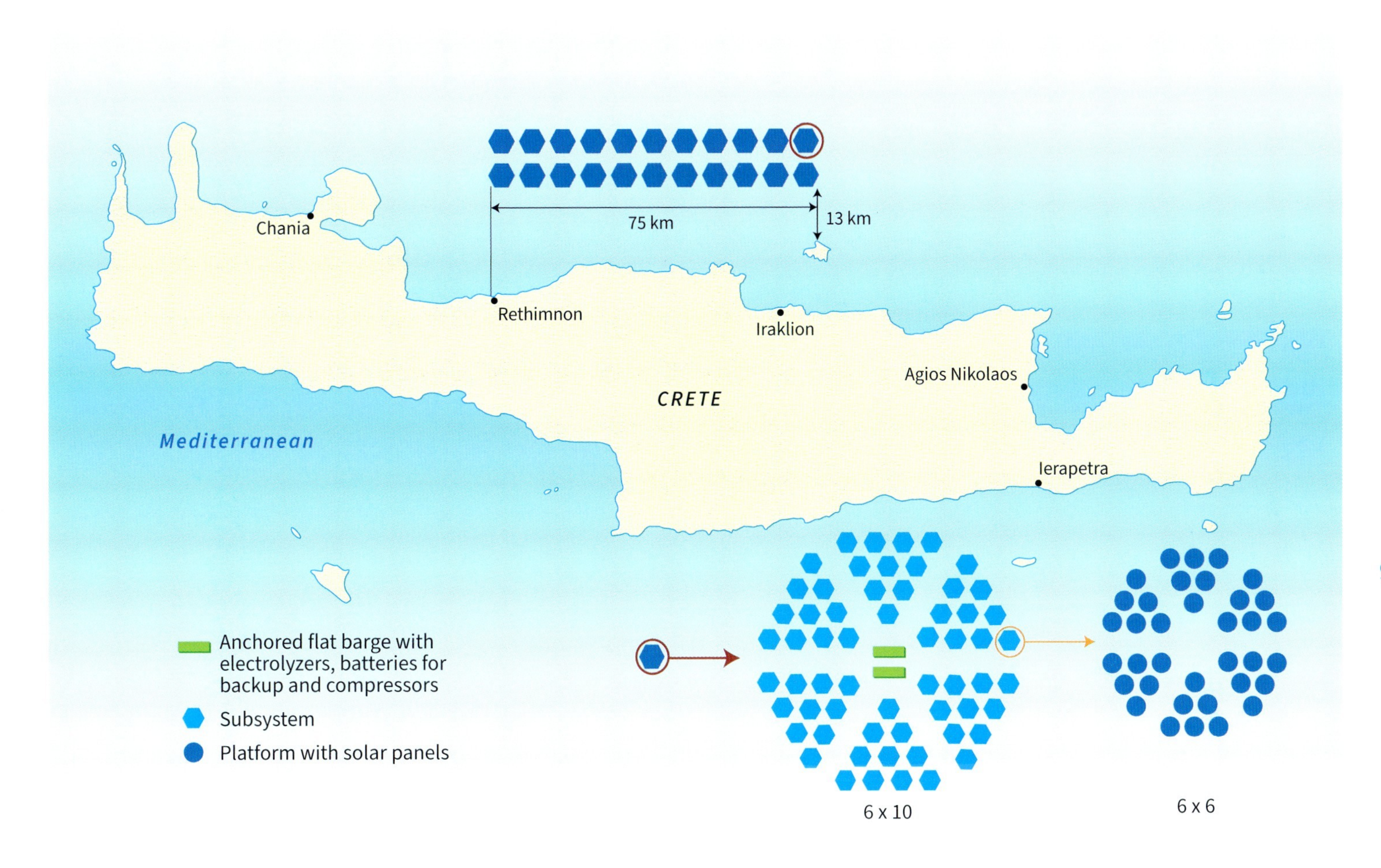

On the Mediterranean Sea we can deploy floating panels with full-load hours of about 1,800. To produce 1 million tonnes of hydrogen, 30 GW of solar panels is needed, together with 21 GW (70% of 30 GW) of electrolyzers. One study was carried out for a floating solar farm off the coast of Crete. The study was based on the deployment of the Ocean Sun floating structure, upon which the solar panels are installed. The electrolyzers, back-up batteries, and compressors are placed on a moored flat barge. This solar-hydrogen system produces 1 million tonnes of hydrogen per year.

A floating offshore solar-hydrogen farm of this kind requires much more solar and electrolyzer capacity than a floating offshore wind-hydrogen production farm. Nonetheless, the hydrogen production costs can be of the same order of magnitude. The investment costs per kW for these solar farms are in fact many times lower than those for wind farms. On the other hand, the spatial requirements of a floating offshore solar farm are many times greater than those of an offshore wind farm.

Energy from temperature differences

Ocean water is a source of energy. You can draw cold water from the sea and use it to cool buildings through seawater air conditioning. And you can exploit the temperature difference between surface water and deeper waters to generate electricity (ocean thermal energy conversion [OTEC]).

Seawater air conditioning

In many tropical areas buildings need to be cooled throughout the year. On islands and in coastal cities, this no longer requires an air conditioner run on electricity, but can also be done by using deep seawater. Seawater air conditioning (SWAC) is feasible for example in the Caribbean, with all its islands, as well as in a number of cities on the coasts of Africa, Asia, Australia, and South America.

SWAC works very simply. You first draw cold water (at about 6 °C) from a depth of about 800 meters to the surface. By means of a heat exchanger, you transfer the cold to a cooling circuit for buildings, similar to the cooling circuits that are currently already used. When the seawater leaves the heat exchanger, it has been warmed up to about 12 °C; you then pump this water back to a depth where the seawater temperature is at this level.

Cold seawater can be used not only to cool buildings, but also to extract freshwater from the air

This form of SWAC can be used in areas that have a high demand for cooling and are situated next to a sea where the coastline drops off very rapidly. Under less ideal conditions, you can also use variations of this system. Seawater with a temperature above 6 °C can be cooled further to the desired temperature by means of a heat pump. Moreover, seawater can also be used as a source of heat for buildings. To this end, you can use a heat pump to further raise the temperature of shallow seawater.

Freshwater production with cold seawater

Cold seawater can be used not only to cool buildings, but also to extract freshwater from the air. This principle is also simple. In hot areas with high air humidity, the water vapor in the air precipitates on cold surfaces. This, for example is what you see happen when you take a cold glass of water out of the refrigerator. This condensation occurs because cold air can hold less water vapor than warm air. The cold water in the glass cools the air around the glass, causing the water in the air to condense: small droplets form on the outside of the glass.

If you let cold water flow through pipes just under or on the ground, then the air around the pipe cools down. This causes condensation, which you can use to irrigate crops. In this manner, you can provide trees, grass, plants, and vegetables with water. You can do this with seawater from 5 °C up to even 12 °C, but also with freshwater that you have cooled with cold seawater and a heat exchanger. An irrigation system of this kind is easy to combine with a SWAC system, which is typically installed in areas where air humidity is high, that is, on the coast.

Seawater air conditioning system for buildings [109]

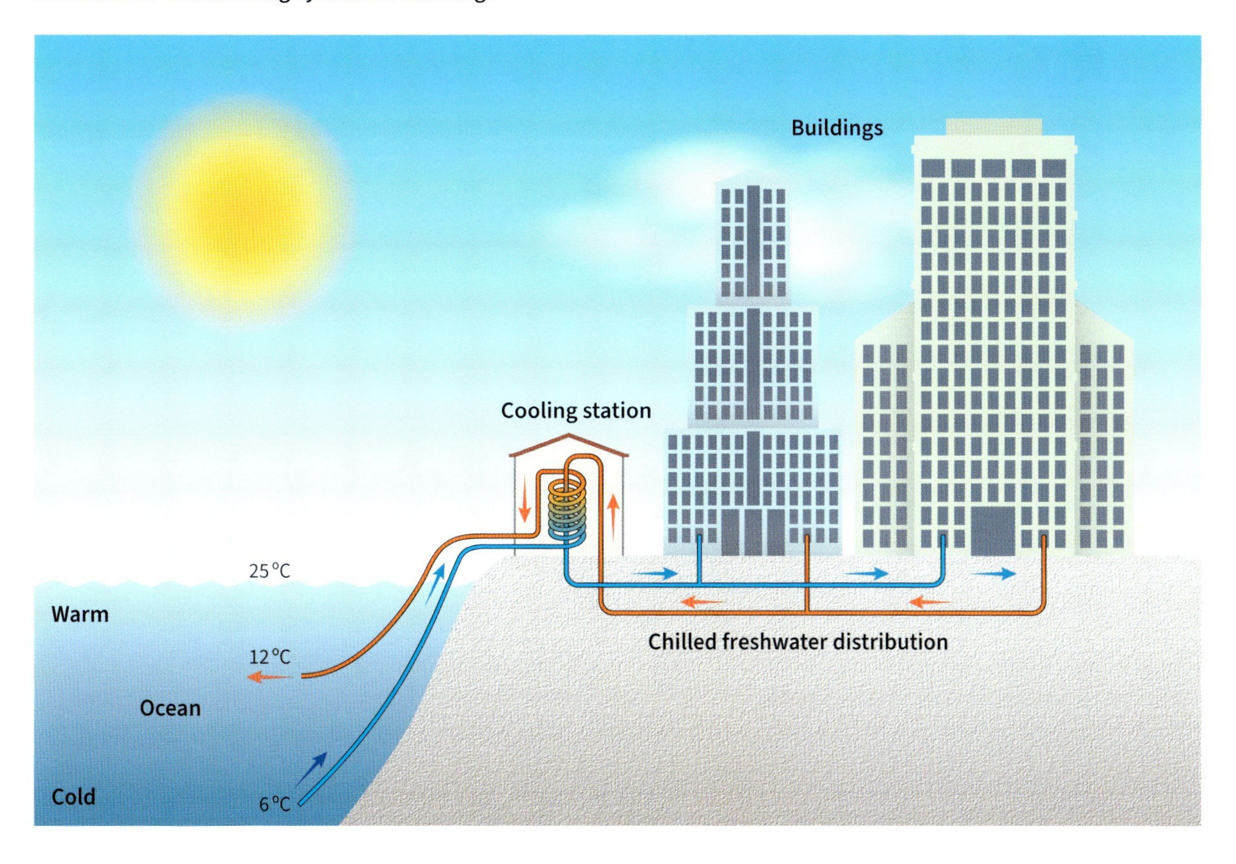

Ocean thermal energy conversion

Of all the solar irradiation that reaches the Earth, 70% shines on the oceans. Most of it warms up the upper layers of the oceans—the largest solar collectors on the planet. The oceans' deeper water is colder. The temperature difference between the surface and deeper water can be used to produce electricity. The water temperature at the surface is 25-30 °C, while at depths from 600 to 1,000 meters it is only 3-7 °C. One finds temperature differences that are big enough for the generation of electricity primarily around the equator, between the two tropics. The ocean thermal energy conversion (OTEC) system has various components, including an evaporator, condenser, turbine, generator, and pump. Electricity is generated as follows. Through heat transfer from the warm surface water in the evaporator, a working fluid with a low boiling point is vaporized. This vapor spins the turbine, which is coupled to a generator that produces electricity. The vapor then flows to the condenser, which is cooled with deep seawater so that the vapor is condensed back into a liquid [110]. A big advantage of OTEC is that it can produce the baseload (continuous flow) electricity. Pilot projects are underway worldwide. The problem however is that an OTEC installation generates relatively little electricity because of the limited temperature difference. An additional, cost-boosting, feature is that they float on the sea, thus in a corrosive environment, usually far away from the coast. OTEC installations therefore produce rather expensive electricity. A possible solution would be to combine OTEC installations with floating wind and/or solar farms.

Energy from tides, waves, and osmosis

Besides wind, sun, and water temperature differences, the oceans offer even more sources of sustainable energy: the tides, the waves, and the transitions from freshwater to saltwater.

Tidal energy

Tides are caused by the gravitational force, and by the rotation of the moon around the Earth and of the Earth around the sun. Tides produce water flows and water level differences (ebb and flow) on the coast, from which you can harvest energy. This is one of the few energy sources that is neither directly nor indirectly derived from solar energy.

There are a variety of ways of harvesting tidal energy. You can use the water flow produced by the ebb and flow of tides to drive underwater turbines. Another option is to fill a barrage at high tide and empty it again at low tide. You can produce electricity with turbines driven by the water flow.
The advantage of tidal energy is that it is very predictable. On the other hand, there are only a few locations where the tide differences or flow speeds are big enough. Moreover, the number of full-load hours is small, with the result that tidal energy plants produce relatively expensive energy.

Wave energy

Waves are also a source of sustainable energy. Waves are created by the winds blowing on the sea surface. As long as waves propagate more slowly than the wind speed above, energy is transferred from the wind to the waves. Wave energy is thus an indirect form of wind energy.
There are different principles and technologies for the harvesting of wave energy, including absorbers, attenuators, over-toppers, oscillating water columns, and undulation. All of them require complex, mechanical installations in or under the water. These technologies can also have an impact on the subsurface marine environment. Nonetheless, wave energy has a reasonable potential. In combination with offshore wind and/or solar farms, wave energy could possibly make a contribution to the generation of sustainable energy.

Osmotic energy

Osmotic energy can be harvested from the flow of freshwater into saltwater. This is energy that becomes available as a result of the difference in salt concentrations between freshwater and saltwater. If you place a membrane in a basin with freshwater on the one side and saltwater on the other, the freshwater will flow through the membrane to the saltwater, until the salt content on both sides is in equilibrium. You can harvest energy from this difference in salt concentration.

The salt concentration in the North Sea is about 35 grammes per liter. In theory, when the salt concentration difference between freshwater and saltwater is 35 grammes per liter, there exists an osmotic pressure difference of 29 bar, that is equal to a height difference of 290 meters. In other words: if 1 cubic meter of freshwater flow into the North Sea, this will theoretically generate as much energy as letting 1 cubic meter of water fall from a height of 290 meters [111].

In theory it would therefore also be possible to develop an osmotic pump that pumps freshwater into the sea from below sea-level, to regulate the water levels and consumes no electricity, but produces net electricity. This could offer interesting possibilities in dams and sea defenses in Bangladesh, the Netherlands, and other low-lying countries on the coast. The harvesting of osmotic energy is however not yet fully developed, particularly because it requires an extensive membrane surface and the membrane can't be allowed to become clogged with dirt.

Wave-energy generator

Opportunities for islands and coastal areas

Interconnected systems can be developed around islands and in coastal areas, where the potential for sustainable energy production is large. In these areas you could also produce chemical products, synthetic fuels, materials, and food using sustainable energy, seaweed, air, and saltwater.

Sustainable energy produced close to the coast is transported to shore as electricity, where it is used to power appliances, light, communication, data centers, transport, and freshwater production.
Sustainable energy produced farther away can be brought to land in the form of hydrogen through pipelines. On land, the hydrogen can be liquefied, transformed into ammonia, or bound to another molecule—a liquid organic hydrogen carrier (LOHC)—so that it can be transported worldwide by ship. Once it is brought to land, hydrogen can also be used as a feedstock for chemical products and synthetic fuels, and used for transport.

Additionally, the seaweed cultivated between the wind turbines can be harvested and shipped. This crop can be used as livestock and fish feed, as a feedstock for pharmaceutical and chemical products, as an energy source, and as a source of carbon for chemical products and synthetic fuels. Freshwater can be made from saltwater, and all sorts of chemical products and materials can be recovered from the brine.

Lastly, you can cool buildings and greenhouses on the coast with deep seawater (SWAC). And, with the same cold seawater you can condensate freshwater from the air, and use it to irrigate trees and plants.

If you realize all these possibilities on a single island, then such an island would sustainably meet its own energy and water needs, as well as export a variety of energy carriers, chemical products, and materials by ship or pipeline.
There are many islands and coastal areas with the potential of developing into such a location for the export of sustainable energy and sustainable products. Examples would include islands in the Caribbean region and the Pacific Ocean, around the Philippines, Madagascar, Sri Lanka, and Hawaii, as well as islands like the Orkneys or Iceland. There are undoubtedly many more suitable locations. It is also conceivable that artificial (floating) islands could be set up at suitable locations, especially for the production and export of sustainable energy, chemical products, synthetic fuels, and materials.

Many islands and coastal areas can begin exporting sustainable energy, chemical products, synthetic fuels, materials, as well as food

With sun, wind, seaweed, and seawater, islands can produce an array of sustainable energy carriers, chemical products, synthetic fuels, materials, and food

5

CLEAN ENERGY AND MATERIALS FROM BIOMASS RESIDUES

Biomass residues can constitute a cost-effective source of green energy and green carbon. Wastewater treatment plants (WWTPs) will in the future become sources of green energy, carbon, raw materials, and clean water. And thanks to seaweed farming, the oceans will become sources of food, (raw) materials, and energy. Moreover, the "sargassum plague", or the excessive growth of seaweed, can be used in our favor.

Indirect source of solar energy

Solar cells perform better than photosynthesis in the conversion of solar energy into a useful energy carrier. Cultivating biomass for energy production therefore does not make sense. At least on land, because cultivating biomass on the sea can certainly be efficient. Sargassum (seaweed) is therefore no longer a plague, but a blessing.

The cycle of life: photosynthesis and combustion [112]

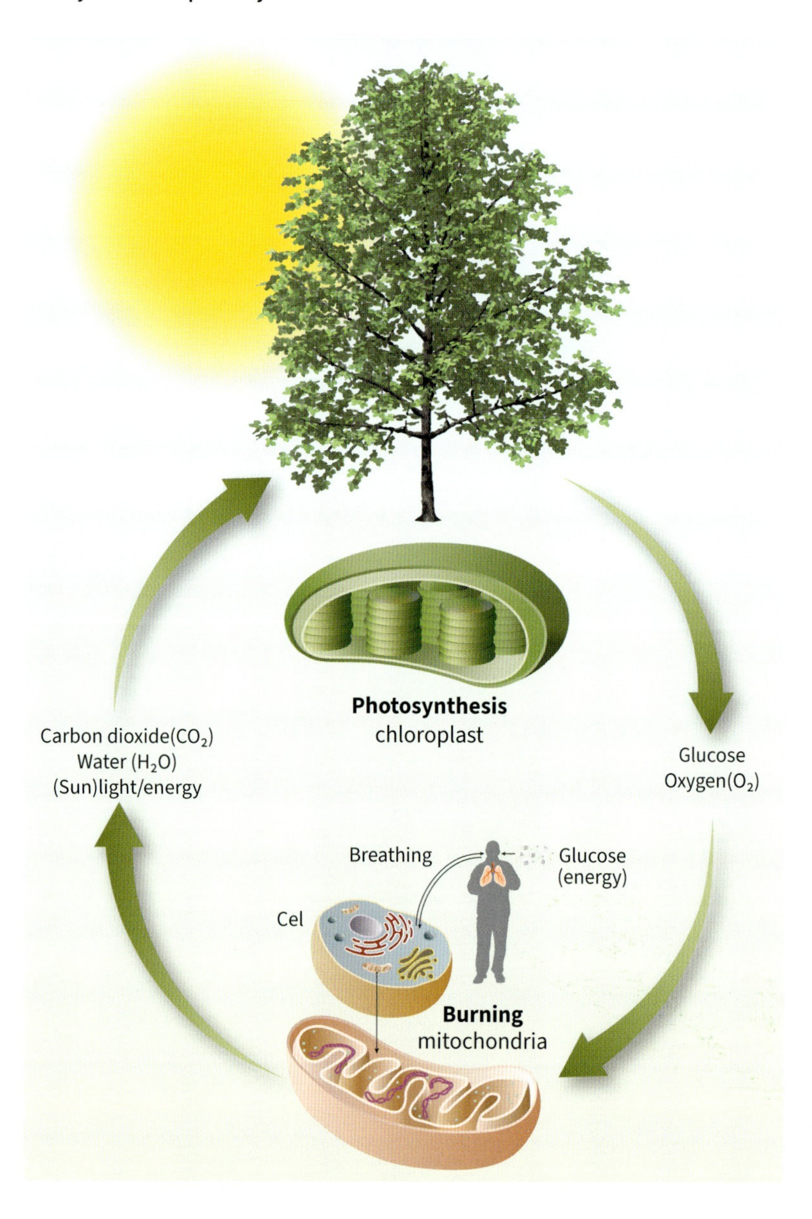

Energy-rich biomass

Besides the sun and the wind, biomass is also an indirect form of solar energy. Plants, trees, and seaweed grow through photosynthesis. Thanks to solar energy, carbon dioxide (CO_2 from the air) and water (H_2O) are transformed into oxygen (O_2) and glucose ($C_6H_{12}O_6$). When animals and humans eat and "burn" this energy-rich biomass, the glucose is broken down with the help of oxygen. The energy it contains is released, together with water and carbon dioxide—the ingredients for plant growth. You may conclude that life on Earth is a cycle, which is powered by the sun.

Food production versus energy production

If biomass contains energy, you could therefore be able to use it as a source of energy. This is not sensible however. Solar panels can actually convert sunlight far more efficiently into usable energy than crops can [113]. Solar panels today have an efficiency of 20-25%, while the maximum efficiency of C3 photosynthesis (the basic process) is 4.6%, and that of C4 photosynthesis (during an intermediary step, a bond is made with 4 C atoms) is 6% at the most [113]. The maximum efficiency, which occurs in corn, sugar beet, sorghum (various grain varieties), and elephant grass, is moreover only possible when sufficient water and nutrients (e.g., nitrogen and phosphate, from manure and fertilizers) are available. Solar panels require none of these. This is why it makes more sense to grow biomass—

Wood production with sawdust as biomass residual stream

certainly on land—for food production than for energy production. It is however clearly useful and sensible to utilize the biomass residues to produce energy. This for example concerns (animal) manure, unusable parts of crops, harvest residue, or residual streams that remain after processing (e.g., wood shavings and treatment sludge). Biomass is also present in the oceans in the form of seaweed, which in some areas even constitutes a plague. Such residual streams can be used not only as energy sources, but most of all as sources of carbon.

Excessive seaweed growth in the oceans

A large amount of nutrients ends up in the oceans because of the runoff of fertilizers. In many places the result is eutrophication, which causes the excessive growth of seaweed. Sargassum, in particular, is a big problem in the Caribbean, and in Central and South America. The seaweed's excessive growth causes considerable health, environmental, and economic damage. But seaweed is also a good raw material. In combination with the production of solar and wind energy, you can use it to supply fully sustainable refineries and chemical complexes with the right energy and feedstocks. In this way you kill two birds with one stone: combat the eutrophication and acidification of the Atlantic Ocean (the “sargassum plague”), and develop a sustainable feedstock and materials system.

It makes more sense to grow biomass on land for food production than for energy production

Source of hydrogen and carbon dioxide

It is better to convert biomass into bio-hydrogen and bio-CO_2 than to combust or digest it. You can use hydrogen as a feedstock and energy carrier, and carbon dioxide for CO_2 dosing in greenhouses and as a feedstock in industry. Carbon is in fact necessary in many chemical products and synthetic fuels.

Green energy and carbon

Residual streams of dry biomass (e.g., straw, sawdust, and wood waste) and of wet biomass (e.g., manure, sludge, and fruit and vegetable waste) currently play an important, often local, role in the generation of sustainable energy. Dry biomass can be incinerated, and the thereby released heat can be used for cooking or heating. It is also possible to make pellets from dry biomass, which can be used to (co-)fire electric power plants. Wet biomass is typically digested into biogas, which can be burned and the generated heat used for cooking or heating. Biogas can also be used in a gas turbine or gas engine, with which to produce electricity and heat. In recent years biogas has also been upgraded to natural gas quality, to then be distributed through the gas network. Or it can be used to make bio-CNG (compressed natural gas) or bio-LNG (liquefied natural gas), for use in vehicles and vessels.

In a sustainable energy system, biomass residues can also serve as a source of green carbon. Carbon is needed in chemical products (e.g., methanol, plastics, and paint) and in high energy-density fuels (e.g., diesel, aviation fuel, and gasoline). Currently, we extract the carbon from crude oil, coal, and natural gas—all "hydrocarbons", compounds of carbon and hydrogen. In a sustainable energy system the carbon will have to come from other sources, such as biomass residues. Or we can recover it from chemical materials at the end of their lifecycle. In all likelihood, the extraction or recovery of carbon from these two sources will be cheaper and simpler than the capture of carbon (in the form of CO_2) from the air (direct air capture) [114].

To bring all of this about, you can convert biomass streams in the future into bio-hydrogen and bio-carbon dioxide. Hydrogen can be used as feedstock and as an energy carrier; carbon dioxide can be used as a feedstock for products, for which we currently still use fossil carbon, and for CO_2 dosing in greenhouses, to stimulate plant growth.

In a sustainable energy system biomass residues are a source of green energy and—especially—of green carbon

Hydrogen transport through the natural gas infrastructure

Biogas that has been upgraded to natural gas quality is now often distributed through the natural gas network. If natural gas activities ceased completely, then the natural gas infrastructure would need to be maintained exclusively for the transport of biogas, the available volumes of which are far smaller. But this same natural gas infrastructure is also suitable for the transport of hydrogen. If you convert the biogas into hydrogen and CO_2, then you solve the problem:

you can use the existing natural gas infrastructure entirely for the transport of hydrogen. You thereby avoid a natural gas lock-in, and you simultaneously accelerate the transition to a sustainable energy system. Ultimately, we would use the natural gas infrastructure exclusively for the transport of hydrogen, from electrolysis, from biomass residues, and, in a transitional period, also still from fossil energy sources. You could also continue distributing biogas from biomass residues via a local network, if this is better suited to local conditions.

In summary, the reasons for the conversion of biomass residues in the future into hydrogen and CO_2 are the following:

- Hydrogen can be used as an energy carrier and feedstock, and CO_2 as feedstock.
- The use of both bio-hydrogen and bio-CO_2 avoids fossil CO_2 emissions.
- The storage and/or long-term locking of bio-CO_2 in products results in negative CO_2 emissions.
- A natural gas lock-in is avoided and the existing natural gas infrastructure can be used entirely for the transport of hydrogen.

Anaerobic digesters at a wastewater treatment plant

Positive business case, now and in the future

Upgrading biogas to natural gas quality and then converting it into hydrogen and CO_2: this involves two conversion steps, each with an associated energy loss and additional investment costs. Why is it nonetheless economically attractive, now and in the future? Because you are making not one, but two products, namely, hydrogen and carbon dioxide. The prices of the two develop in different ways. Let's first look at the technology. After biogas is upgraded to natural gas quality, we convert it into H_2 and CO_2. This is done with a small-scale steam methane reformer, or with an autothermal reformer. If you use manure or another plant waste as the biomass residual stream, then for every kilo of hydrogen you will also produce about 15 kilos of CO_2 [115]. The hydrogen is transported through a pipeline to the energy system, where its price needs to be competitive with hydrogen from other sources. We liquefy the CO_2 and transport it by road to clients, such as a chemical factory, a biofuel plant, or a greenhouse horticultural area.

Production costs and income for 1 kg of hydrogen plus 15 kg of carbon dioxide from biogas, now and in the future [115]

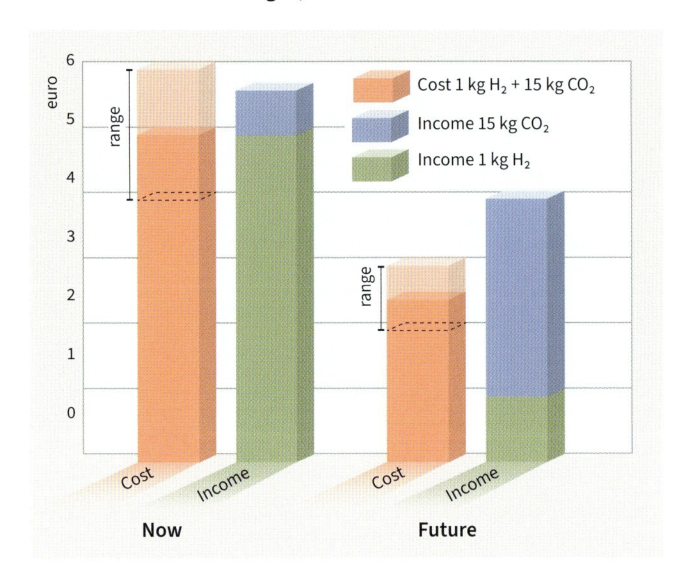

How do you make hydrogen and carbon dioxide from biomass residues?

You can use gasification to convert dry biomass residues and plastic waste into a syngas gas, which consists of methane (CH_4), hydrogen (H_2), carbon monoxide (CO), carbon dioxide (CO_2), and nitrogen (N_2). You can then further convert and separate this syngas into H_2, CO_2, and N_2.
Wet biomass can first be digested, which produces a biogas consisting mainly of CH_4 and CO_2, or gasified, which produces a bio syngas consisting of H_2, CH_4, CO, and CO_2. You can then convert this biogas or bio syngas into H_2 and CO_2. There are various gasification and digestion technologies for the production of a bio syngas or biogas. The technologies needed for the subsequent production of hydrogen and carbon dioxide are the same ones that use fossil fuels (coal, oil, or natural gas) to make hydrogen and CO_2.

This conversion of biogas into H_2 and CO_2 is economically interesting. The price of green hydrogen will decrease in the future because the electrolysis of water with sustainable energy will become cheaper. For its part, the price of green CO_2 will increase because CO_2 will become scarcer. In light of these contrasting price developments, a positive business case is always possible, now and in the future. We can illustrate this with a scenario for the price development of hydrogen and CO_2. Today, the cost of producing 1 kg of hydrogen plus 15 kg of CO_2 is 4-6 euros. Technological and efficiency improvements will reduce these costs in the future to 2-3 euros. The income is a function of the hydrogen price (let's assume this is now about 5 euros a kilo, and 1 euro a kilo in the future), and of the CO_2 price (let's assume this is now 50 euros a tonne, and 200 euros a tonne in the future).

Integrated wastewater treatment

Wastewater treatment plants process a large biomass residual stream: the wastewater of citizens and companies. In the past this stream was only treated, but today energy and raw materials are also recovered from it. In the future a further step will be taken: the recovery of clean water and the production of green carbon.

How to treat wastewater?

The first, primary treatment, stage in a wastewater treatment plant (WWTP) consists of the removal of large solids from the wastewater by means of filtration and sedimentation. Then, in the secondary treatment stage, bacteria break down the waste material in a process accelerated by aerating the wastewater. After a period of up to a few days, the wastewater has been "picked clean". The treated water is separated from the "activated sludge", a mixture of dirty water and sated bacteria. It then undergoes tertiary treatment—disinfection, filtration, and sometimes also nutrient removal—and is then discharged into surface water. Part of the activated sludge is reused in the treatment of new wastewater, and a significant part is left over. You can use this surplus sludge to produce biogas, but it can also be incinerated or landfilled. Following the production of biogas, the digested sludge is incinerated or used as a fertilizer in agriculture.

New technologies are becoming more and more effective, using light or air, in increasing the yield of the sludge—and thereby the biogas production. The further development of these technologies and their large-scale application are however uncertain.

Energy use in wastewater treatment

Wastewater treatment plants use energy mostly in the form of electricity, for purposes including aeration, pumping, and sludge dewatering. Furthermore, the WWTP also needs energy to heat buildings, maintain the digester at temperature, and to transport the left-over sludge with trucks. The average electricity use is about 0.5 kWh per m^3 of wastewater, with a range from 0.2 to 0.9 kWh/m^3 [116], [117]. The average daily water consumption in the EU is 150 liters per person [118]. Practically all of this water, about 55 m^3 per person annually, flows through the sewer system to a WWTP. This is exclusive of stormwater, which is also often transported through the sewers. The treatment of all of this wastewater therefore requires approximately 28 kWh per person annually. This represents 1-2% of the domestic electricity use per person, that is, about 1,600 kWh per person annually [119]. Aeration accounts for more than half (55-70%) of a WWTP's electricity use, pumping of wastewater streams consumes about 15%, while sludge dewatering accounts for about 7% [120].

Laughing gas emission by wastewater treatment

Laughing gas, or nitrous oxide (N_2O), which is a greenhouse gas that is 265 times as powerful as CO_2, forms in wastewater. After CO_2 and methane, laughing gas is the biggest contributor to greenhouse gas emissions. Wastewater contains on average about 0.05 kg/m^3 of nitrogen, in the form of ammonia, which comes from urine and fecal matter [121]. A WWTP produces about 0.03 kg of laughing gas per kg of nitrogen [122]. That is 1.5 kg of laughing gas per 1,000 m^3 of wastewater. This represents a greenhouse gas equivalent of almost 400 kg of CO_2. Every year, WWTPs in the Netherlands and in Europe emit, respectively, 310,000 and 15,500,000 tonnes of CO_2 equivalent in laughing gas.

Useful application of oxygen and heat

Efforts to reduce WWTP energy use and CO_2 emissions are mainly targeted at saving energy, increasing the biogas yield, and using sustainable electricity. A broader system approach could achieve even more. Such an approach begins with the production of hydrogen. You make hydrogen through the electrolysis of water in an electrolyzer that is fed by a portion of the (self-generated and purchased) sustainable electricity. But the production of hydrogen through electrolysis also generates oxygen. And if you use this oxygen instead of air for the aeration, then you have to pump only one-fifth of the volume of air through the water, because the bacteria only need oxygen. You therefore make a considerable saving in the aeration's electricity use. And because you use oxygen instead of air, the laughing gas emissions are also lower [124]. Lastly, you can use the residual heat from the electrolyzer to maintain the digester at temperature. In this way you end up having more biogas—the biogas that you would otherwise have used for this heating.

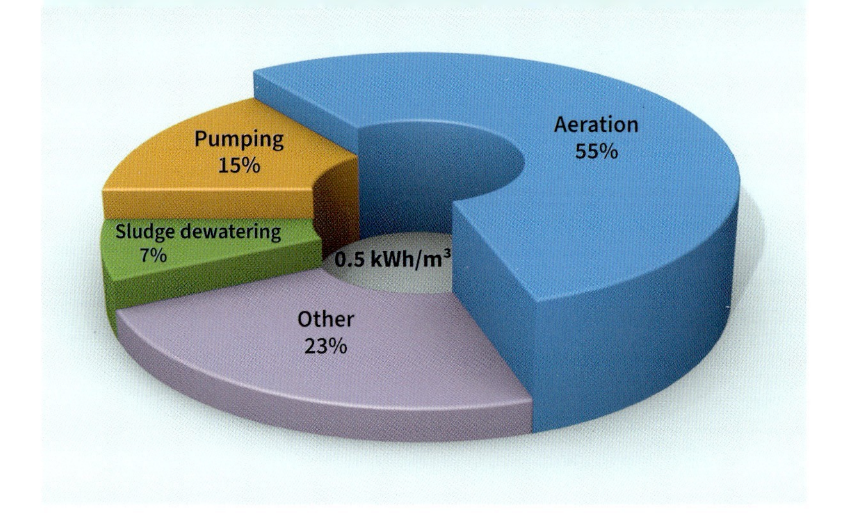

Electricity use in wastewater treatment plants

At wastewater treatment plants, one can make good use not only of hydrogen, but also of an electrolyzer's oxygen and residual heat

Laughing gas production and emissions at wastewater treatment plants in the Netherlands and Europe

	Nederland	Europa
Laughing gas production per 1,000 m^3 of wastewater	1.5 kg	1.5 kg
Water use per person	120 liters/day [123]	150 liters/day [118]
Inhabitants	17.7 million	710 million
Laughing gas emission	1,200 tonnes/year	58,300 tonnes/year
CO_2 equivalent	310,000 tonnes/year	15,500,000 tonnes/year

How much useful heat does an electrolyzer produce?

The amount of heat produced by an electrolyzer corresponds to the energy loss that occurs in the conversion of electricity into hydrogen. An electrolyzer with an HHV efficiency of 80% produces 80 kWh of hydrogen from 100 kWh of electricity. The remaining 20 kWh consists of heat (at 60-80 °C). Because this heat is located in different parts of the system, in the electrolyzer itself, the hydrogen stream, and the oxygen stream, you can't use all of it—probably 10-15 kWh [125].

For the aeration an electrolyzer is needed that continuously supplies the same amount of oxygen and heat. This won't work if the electrolyzer is only connected to wind and/or solar power. A connection with the electric grid is therefore also necessary. This can be done in a smart manner, that is, by connecting half of the installed solar and wind capacity to the electric grid, and the other half to the electrolyzer. With the same electric grid capacity, you can now install and utilize twice as much solar and wind capacity. Moreover, you can run the electrolyzer continuously at full capacity; the connection capacity of the electric grid is after all as large as the capacity of the electrolyzer. Part of the electricity required comes directly from the solar and wind capability, and the other part you purchase as green power and is transmitted through the electric grid.

Biogas conversion into H_2 and CO_2

Combusting biogas in a gas engine, as is currently still often done, is inefficient because you can't make good use of most of the heat. This is why biogas is increasingly being upgraded to natural gas quality, so that it can be fed into the natural gas network. If you are going to produce hydrogen you will therefore need to install a new (separate) pipeline. This investment is not necessary if you convert the biogas into hydrogen and CO_2: you can then always repurpose the natural gas pipeline as a hydrogen pipeline for the hydrogen from both the electrolyzer and from the biogas.

TECHNOLOGY FOR THE FUTURE

Biopolymers

Bacteria make raw material for bioplastics from treatment sludge

Bacteria degrade biomass residues into small fragments (molecules), like CO_2 and H_2. With chemical technology we can use these fragments to make all kinds of organic substances, known as synthetic polymers. Under specific culture conditions, bacteria can themselves also make polymers. These are known as biopolymers. After many years of investigation, researchers have succeeded in getting bacteria to produce biopolymers from treatment sludge [126]. All you need to do is extract the biopolymers from the bacteria and you'll have the raw material for bioplastics. An additional benefit is that these bioplastics are biologically degradable. Through bacterial fermentation (digestion) you can for instance make polyhydroxyalkanoates (PHAs), a very multifaceted raw material, with which you can for example make degradable packaging and grow-pots for horticulture, or use in the 3D printing of medical bio-implants. An example of such an implant is a stent, which is used to keep blood vessels open. PHA applications also make it possible to dose medications very precisely and deliver them to the right place in the body [127].

Medical application of PHA: a 3D-printed stent (1 mm diameter) to keep blood vessels open [128]

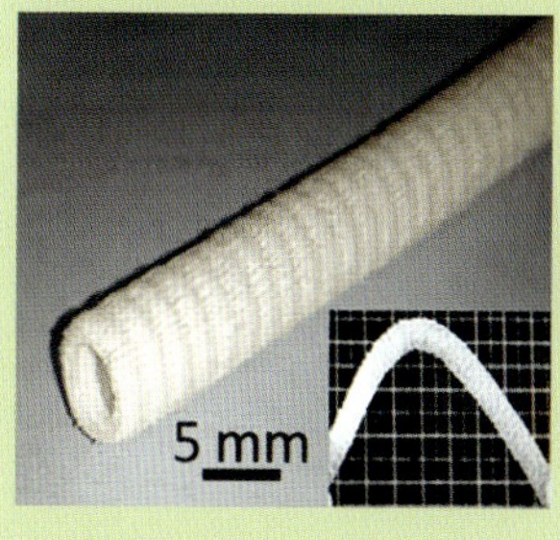

Furthermore, you can make good use of the generated CO_2, for instance in horticultural greenhouses. If there is a greenhouse area in the vicinity, you can use a pipeline to transport the CO_2 to it. Alternatively, you can also liquefy the CO_2, and then transport it by (hydrogen-fueled) tank truck to a chemical company or synthetic fuel producer, where it will be used as a feedstock. Carbon dioxide can be liquefied with different combinations of pressure and temperature. In tank trucks it is transported at a pressure of about 20 bar and a temperature of -30 °C.

Ammonia recovery from wastewater

Wastewater contains a great deal of nitrogen in the form of ammonia. The discharge of ammonia into surface water is not desirable because it leads to algal blooms and eutrophication. For this reason it is converted into nitrogen in the WWTP. This process uses lots of energy. But besides the high energy consumption, the loss of the ammonia is also a shame, since it could also actually serve as a fertilizer and energy carrier. The production of fertilizer uses a lot of energy. The Haber-Bosch process, which produces ammonia (fertilizer) from nitrogen from the air and hydrogen, accounts for 1-2% of global energy use [129]. This fertilizer is applied to the land to produce plant proteins, and part of it is used for livestock feed. The livestock then transform the feed into animal proteins. Because of all the losses in the production chain, only 17% of the produced nitrogen is captured in proteins for human consumption [129]. The conversion of proteins in our bodies produces ammonia, which ends up in the sewer through our urine. It can be advantageous, from an energy and economic perspective, not to destroy this ammonia in a WWTP, but rather to directly recover it from the wastewater as fertilizer—or perhaps, in the future, as protein. The conversion of ammonia into nitrogen in fact requires the same amount of energy as the production of ammonia from nitrogen and hydrogen using the Haber-Bosch process, that is, 12.5 kWh/kg of nitrogen [129].

Currently, it is (still) best to recover ammonia from the watery residual stream that flows out of the digester. This stream actually contains high concentrations of ammonia, and there are a variety of techniques to capture it—for instance, as ammonium sulfate, a product that can be used as a fertilizer.
In the future you will also be able to use aerobic hydrogen-oxidizing bacteria to directly convert ammonium sulfate into proteins. This is done by feeding the bacteria with ammonia, hydrogen, and CO_2 [130]. By producing proteins directly, you no longer need to go through the very inefficient process of growing plants with fertilizer, water, and solar energy, in order to then extract proteins. This inefficiency is all the greater if you feed animals with corn or soja, to then extract animal proteins.

The potential for ammonia recovery is large. On an annual basis, it amounts to about 1.5 kg of "recoverable" ammonia per person, based on the average volume of wastewater [131]. If we recover all of this ammonia and reuse it as fertilizer, we would save 110 billion kWh of energy globally. If this direct use of recovered ammonia were to encounter resistance, such as legal obstacles or lack of social acceptance, it could still also be converted again into hydrogen.

Reuse of treated wastewater

All over the world treated wastewater is being discharged into surface water or into the sea. This treated water is also usable for irrigation, and you could also use it to make demineralized water or drinking water. It is a shame to let this water drain away, all the more so given that many parts of the world are affected by freshwater shortages. The production of hydrogen and oxygen through electrolysis requires very pure, demineralized water. To make this water, you first need to pre-treat and then demineralize it using a reverse osmosis (RO) installation. The result is water of high quality, which you can make into drinking water by means of a simple post-treatment process.

From an energy and economic perspective, it is advantageous to recover ammonia directly from wastewater

Integrated system for wastewater treatment and sustainable energy production

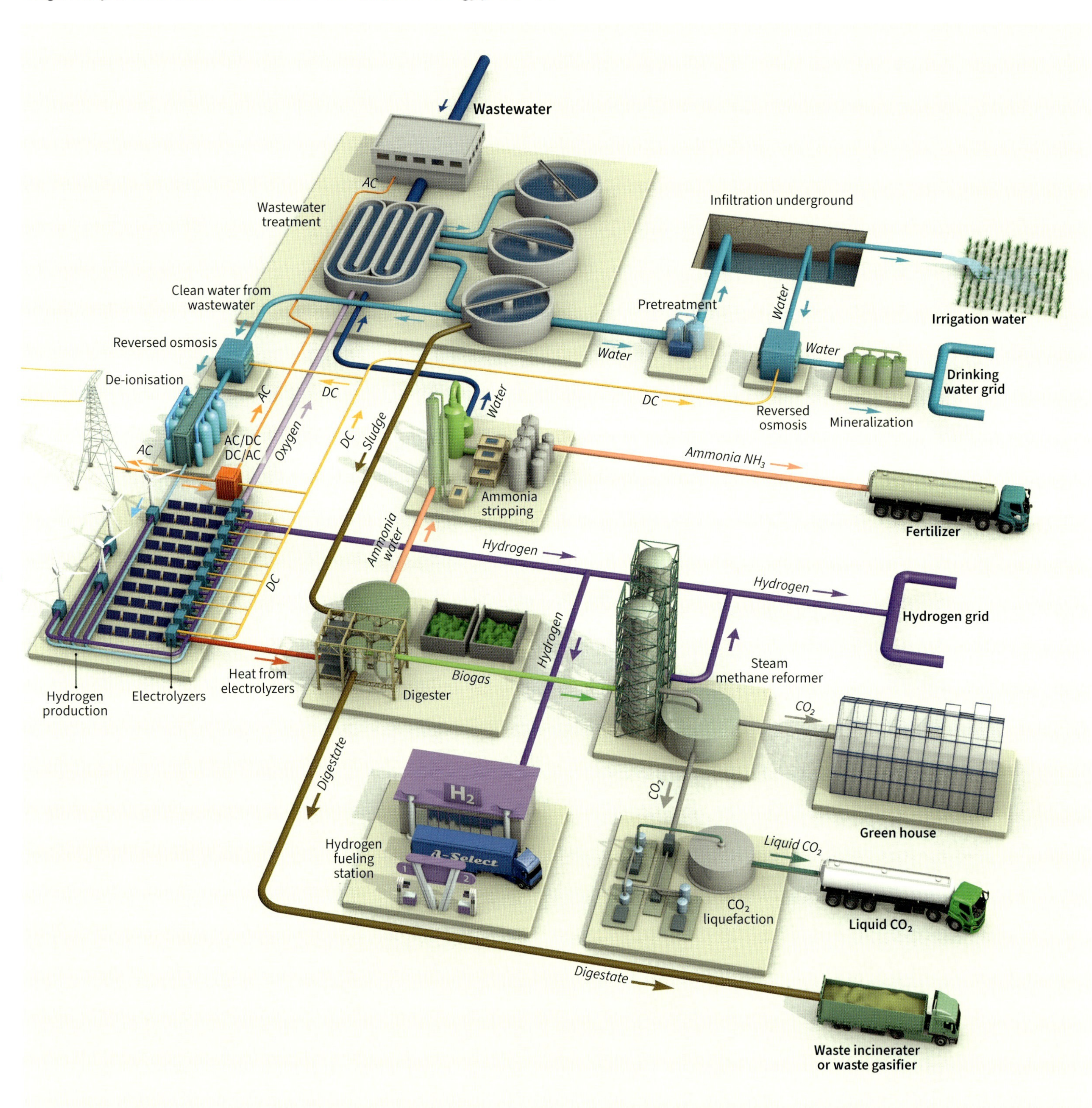

To make it suitable for the electrolyzer, the water needs to be made even purer. This is done through a de-ionization process.
To produce water for hydrogen you only need a small RO installation. If you use a larger one, then you can also make drinking water. You could first store the treated wastewater underground, where the water would be further purified by means of ground passage. Following reverse osmosis (and remineralization), drinking water can then be delivered. Naturally, to prevent any health risks, a quality control needs to be done before this water is returned to the drinking water network.

Integrated wastewater treatment

The combination of wastewater treatment with sustainable energy production, creates an integrated system that produces green hydrogen, green CO_2, ammonia, irrigation water, and drinking water. In the future, such a system could even produce bioplastics and proteins, with no CO_2 emissions, and with fewer to absolutely no emissions of methane and laughing gas. The most important benefits of such an integrated system are the following:

1. Production of green energy from the sun and wind.
2. Production of green hydrogen through the electrolysis of water with renewable electricity.
3. Possibility of integrating two times as much solar and wind power into the energy system through a smart combination of electrolyzer capacity and electric grid connection electricity.
4. Production of green hydrogen and green CO_2 through biogas reforming.
5. Good use of electrolyzer oxygen, thereby saving 50-75% on electricity use for aeration [124].
6. Heating of the digester with residual heat from the electrolyzer.
7. Lower methane and laughing gas emissions.
8. Recovery of ammonia from wastewater, thereby saving on the use of electricity and hydrogen for the production of new ammonia.
9. Besides the production of demineralized water for the electrolyzer, the production of irrigation water and potentially of drinking water by increasing the size of the reverse osmosis installation.
10. Temporary underground storage of treated wastewater for use in dry periods and as an additional treatment stage.

Sargassum: from plague to useful raw material

The *sargassum* seaweed has become a "plague" in many areas because of its excessive growth. It washes up in huge quantities on the coast, where it causes health, environmental, and economic damage. Here we will show how these problems can be solved, while the potential of seaweed, as a source of food, materials, and energy, can be simultaneously harnessed.

Excessive seaweed growth in the oceans

Large quantities of nutrients are ending up in the oceans because of the runoff of fertilizers. In many areas this causes eutrophication, which then leads to the excessive growth of seaweed. Sargassum, in particular, is a big problem in the Caribbean, and in Central and South America. Excessive growth of this brown algae causes considerable health, environmental, and economic damage. But seaweed is also a good raw material. In combination with the production of solar and wind energy, you can use it to supply fully sustainable refineries and chemical complexes with the right energy and feedstocks. In this way you kill

Renewable seaweed refinery-chemical complex, with seaweed and water as raw materials, and solar and wind as energy sources [132]

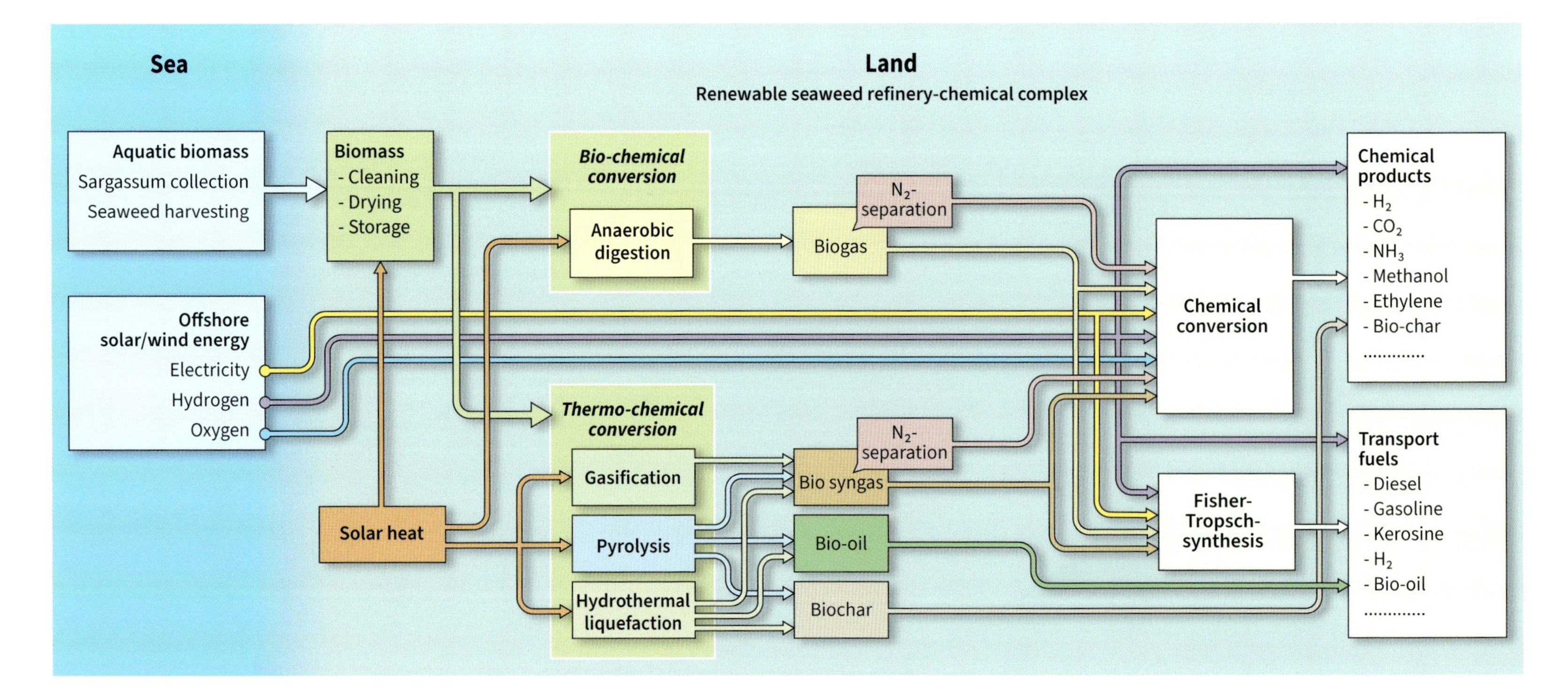

Seaweed farming and harvest

two birds with one stone: combat the eutrophication and acidification of the Atlantic Ocean (the "sargassum plague"), and develop a sustainable feedstocks and materials system.

Seaweed as food

Fresh seaweed is a valuable material, from which you can make food and other products, and extract energy. Seaweed can be divided into three categories: green, brown, and red, each of which has a specific application as a food. Seaweed is used in various food products (e.g., soup, sushi, wraps, salads), in the production of hydrocolloids (alginate, agar, and carrageenan) which are used as binding agents, and as animal feed and mussel feed. Seaweed is also a source of proteins and oils, with which you can make many health and medicinal products. Seaweed farming is simple. It grows on lines in the water and can be harvested several times every year.

Seaweed production up to 2050

Food production from seaweed is seen as an important part of the solution to food shortage in the world [100]. Seaweed production is rapidly increasing globally. About 12 million tonnes of seaweed was harvested in 2000; in 2019 the figure had tripled to almost 36 million tonnes, the largest part of which was cultivated seaweed. Asia represents 97% of the market, but seaweed consumption is growing fast in other parts of the world as well. In 2019, the size of the world market for seaweed was about 8.4 billion dollars [135].

The production of large volumes of seaweed for human and animal food could dramatically impact food security. According to a World Bank scenario, the growth of seaweed farming will increase annually by 9-14% until 2050. This means that in 2050 about 500 million tonnes (dry weight) of seaweed will be produced. That is approximately 10%

Coastal areas where sargassum (brown algae) washes up [132]

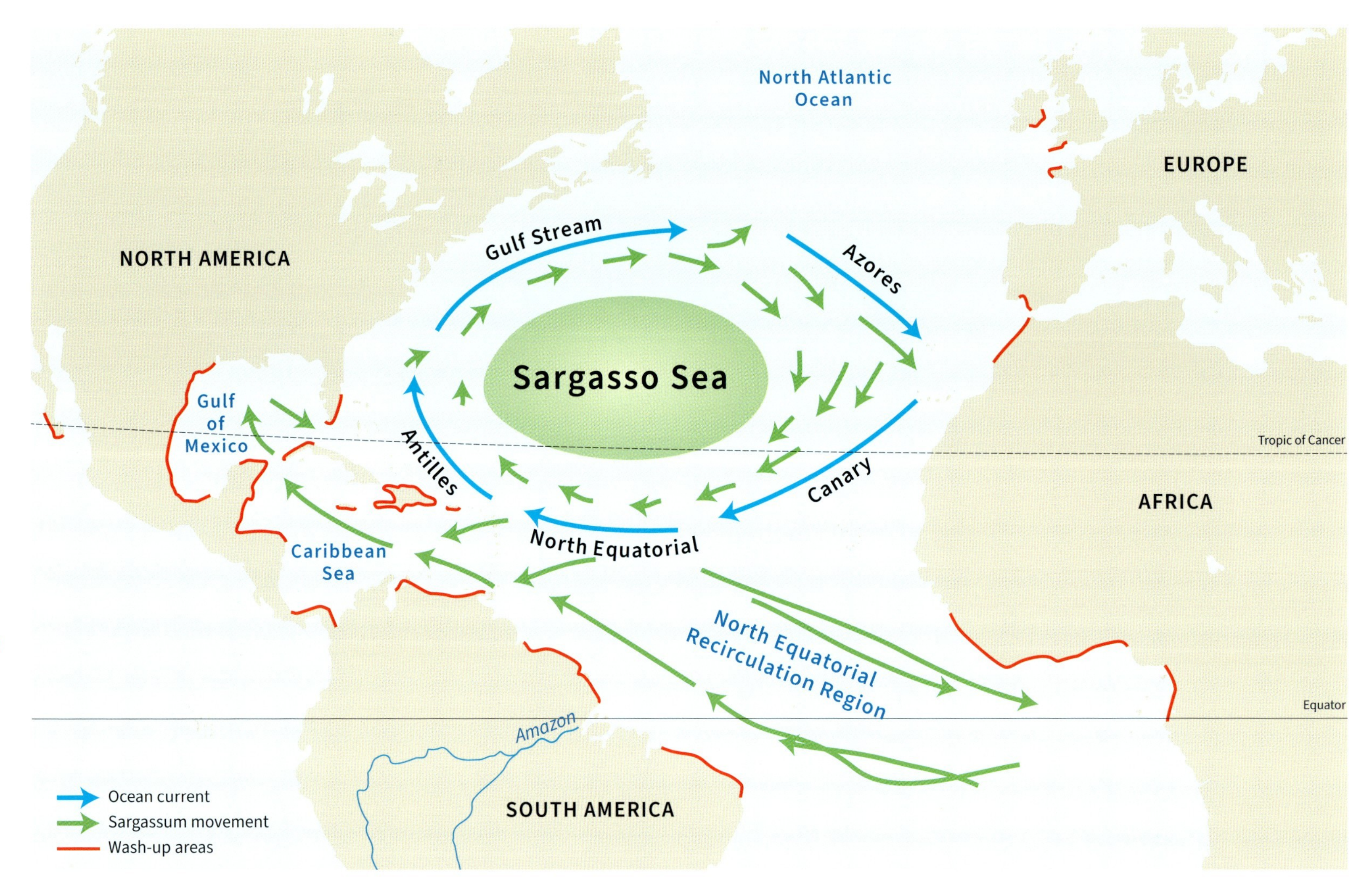

of the current world food supply. Seaweed production is in addition a source of income for many people, including fishermen.

For this scenario, the World Bank assumes an average productivity of about 1,000 tonnes (dry weight) of seaweed per km^2 (10 tonnes per hectare, or 1 kg per m^2)—by its own admission, this is a conservative estimate, which today applies to the best-performing, modern "sea farms". For the total harvest of 500 million tonnes in 2050, a sea surface of about 500,000 square kilometers would then be needed, the equivalent of 0.14% of the surface of all oceans [100].

Less eutrophication and acidification

Today's agriculture uses large amounts of fertilizer to achieve high levels of food production per hectare. It is estimated that 15-30% of these fertilizers ultimately end up in coastal waters, which become "dead zones" as a result [136]. This is because the fertilizer constitutes extra food (nutrients, particularly nitrates and phosphates), which leads to massive seaweed growth in certain parts of the sea. When this seaweed dies, the bacteria that degrade the algae consume so much oxygen that none remains for other life-forms, such as fish.

With seaweed farming such a surplus of nutrients can actually be absorbed. If the seaweed is harvested as opposed to letting it die, these “dead zones” can be brought back to life.

With the production of the 500 million tonnes of dry seaweed in the World Bank’s scenario, about 10 million tonnes of nitrogen would be absorbed. This is approximately 30% of the amount of nitrogen that currently ends up in the sea every year. Besides ammonia or nitrate, phosphate also ends up in the sea. Because world phosphate reserves are limited and depleting rapidly, seaweed could be used as an alternative source of phosphate.

The growing concentration of CO_2 in the atmosphere contributes to the acidification of the oceans, which impacts marine life. The production of 500 million tonnes of dry seaweed in the World Bank’s scenario could absorb 135 million tonnes of CO_2. That is only a small part of the 34.9 billion tonnes of global CO_2 emissions in 2021 [103]. Yet, seaweed can control acidification locally, because it raises the pH (acidity level) of the water (a higher pH value means lower acidity) [137].

Sargassum as plague

Sargassum is a genus of brown seaweeds, comprised of more than 300 different species, each with a specific composition. In the Caribbean, the Gulf of Mexico, and in the area where the Amazon flows into the sea, we see excessive growth of this brown algae. Excessive growth of seaweed, and not only of sargassum, occurs in other parts of the world as well. Large quantities of sargassum wash ashore on the coasts of the Caribbean islands, the Gulf of Mexico, Brazil, Africa, and Europe, causing health, environmental, and economic problems.

The amount of sargassum that washes up on the coast varies by year, season, and location. It is also a function of the ocean currents, the weather conditions (particularly the wind direction), and the coastal morphology (seabed relief, coastline, and dunes). In some months no sargassum washes up at all, but in peak months millions of tonnes can wash up. On beaches, in harbors, and in river estuaries, washed-up seaweed dies, starts rotting, and releases awful odors. This has a serious negative impact on the health, living conditions, and ecosystems on the coast, and on coastal marine life. It is also damaging to economic activities, such as fisheries, tourism, and shipping [138].

Sargassum as energy and carbon source

Fresh seaweed is a valuable material, from which you can make food and other products, and extract energy. Not all seaweed species are equally suitable for all applications and products. Washed-up seaweed always consists of a variable mix of different species. It is also often contaminated with sand and other impurities, or has already begun to decompose. This makes it difficult to use it for food, materials, or energy production. In many areas the washed-up seaweed is therefore collected, but then burned. Those who grow seaweed themselves don’t have these problems, and can specifically choose a particular species or mix of species, depending on the intended application.

Those who grow seaweed themselves, can specifically choose a particular species or mix of species

To prevent sargassum from washing ashore, it has to be collected at sea. This can be done, for instance, by installing a barrier in the water. Seaweed collected in the sea is fresh and not soiled with sand. Of course, the species are still mixed and sometimes contaminated with heavy metals and other substances. Such a mix is not appropriate for food production, but it can be used to make products and for the production of energy.

Products one could make from 1 tonne of wet weight sargassum [139]

1 tonne fresh sargassum	
22 kg	Alginates
80 kg	Activated carbon
811 kWh	Energy
841 L	Fertilizer
2,220	Pairs of shoes
4 tons	Compost (co-composting)
500	Notebooks
44,444	Soap bars
5,000	Drink coasters or business cards
114 kg	Mushrooms
4,926	Biodegradable single-use plates
108	Sargablocks
11	Seabales
99	Sheep receiving 1 week of seaweed supplement in their feed
10 kg	Fucoidan
< 5 kg	Bioactive secondary metabolites

Cleaning up washed-up sargassum is labor-intensive and dirty work [139]

Sargassum barriers at solar and wind farms

In order to exploit the sargassum that accumulates on the sea, you need to collect it at a barrier and then transport it. Barriers of this sort have been installed in a number of places in the Caribbean, primarily in tourist areas. A good example is the 4.2 km-long, floating barrier which has been in place off Punta Cana in the Dominican Republic since 2018. Specialized barges can collect between 50 and 500 tonnes of sargassum per day at these barriers.

As discussed earlier, seaweed farming in the Caribbean can be effectively combined with floating wind farms. But if you grow a specific species of seaweed in a wind farm, you won't want it to be "contaminated" with other species. And no sargassum can thus be allowed to settle on cables and pipelines in floating solar and wind farms. By setting up sargassum barriers around and in between different floating wind and solar farms, you could keep out the sargassum that grows in the wild. This sargassum will therefore not be able to contaminate the cultivated seaweed, nor will it wash ashore. The harvesting of cultivated seaweed could possibly also be combined with the collection of the seaweed caught in the barrier. In this way, sargassum can serve as a source of food, feedstock, materials, as well as energy. And the "sargassum plague" can be transformed into a “sargassum blessing”.

Refinery with chemical complex

If you combine the production of sustainable electricity, hydrogen, and oxygen from floating solar and wind farms, with seaweed farming in wind farms, and the collection of sargassum with barriers around and between floating solar and wind farms, then you will have all the feedstocks you need for a fully sustainable refinery-chemical complex.

Collected and cultivated seaweed needs to be cleaned, dried, and stored before use. A solar collector can be used to heat air to a high temperature. By pumping this air through the seaweed, large amounts of seaweed can be dried at once. This is faster than doing it in the open air. The solar heat can also be used to maintain the digester—to produce biogas from seaweed—at the right temperature, and to further dry the seaweed, so that it can be gasified or pyrolyzed.
If the collection and cultivation of sargassum are combined, there is more control over the quality of the ultimate feedstock. Moreover, there is no need to also use cultivated seaweed as well. However, since the supply of sargassum can vary greatly over time, large storage systems are needed. If using collected seaweed, it needs to be cleaned more thoroughly and the processes need to be designed in a more robust way.

Ultimately the oil- and gas-based refineries in the Caribbean, and in South and Central America—but also at many other places in the world—can be replaced with fully sustainable refineries and chemical complexes, whose feedstocks come from seaweed and water, and whose energy comes from the sun, wind, and seaweed.

You can ultimately replace oil- and gasbased refineries with a sustainable refinery-chemical complex based on feedstocks and energy from the sun, wind, seaweed, and water

What you can make from seaweed in a chemical complex

Through a Fischer-Tropsch process you can make sustainable fuels, like gasoline, diesel, and aviation fuel. This requires biogas (CH_4, CO_2) or a bio syngas (a combination gas of H_2, CH_4, CO, and CO_2, which also contains nitrogen [N_2]), and hydrogen (H_2) from sustainable energy. Through steam methane reforming or autothermal reforming, biogas or bio syngas can be entirely converted into hydrogen (H_2) and carbon dioxide (CO_2). Autothermal reforming requires pure oxygen (O_2), which is produced through the electrolysis of water. Sargassum also contains lots of nitrogen. Together with hydrogen, through the Haber-Bosch process, this can be converted into ammonia, a primary component of fertilizer. From the bio syngas, or from H_2 and CO_2, we can make methanol. And we can make ethylene (C_2H_4) from biogas or naphtha (a by-product of the Fischer-Tropsch process), or even from CO_2 and hydrogen, which is the basic compound used in making polyethylene or a plastic. And there are many more chemical products that can be made using the biogas, bio syngas, hydrogen, oxygen, and nitrogen feedstocks.
We can produce biogas from seaweed using a biochemical conversion route: anaerobic digestion. This is a technology that we have applied for a long time.
We produce bio syngas from seaweed using a thermochemical route, such as gasification, pyrolysis, or hydrothermal liquefaction. The application of the last two processes generates biochar, which is primarily carbon, and bio-oil.

6

TOWARD A SUSTAINABLE ENERGY SYSTEM IN 2100

Solar and wind energy production has become very cheap in a short period of time. A fully sustainable energy system however requires a new design, based on the specific characteristics of sustainable energy sources and carbon-free energy carriers. But how do we move from today's fossil energy system toward a sustainable energy system? And what does an energy system with Sustainable Energy System Goals look like?

From fossil to sustainable

With solar cells and wind turbines we can now produce green electricity for 1 to 2 eurocents per kWh. But this can only be done at a very great distance from the energy demand. How do we now get this cheap sustainable energy in a right form, at the desired time, to the correct location? And how do we move as fast as possible from a fossil to a sustainable energy system?

Emission-free end-use

It is important that the end-use of energy no longer produces any greenhouse gas emissions, particularly of CO_2. This is now already the case with the use of electricity for lighting or appliances. But it is not the case in the use of natural gas, for instance, to produce steam in an industrial steam boiler, or hot water in a heating boiler at home. And the diesel and gasoline in your car's internal combustion engine also produce CO_2 emissions.
The use of natural gas, gasoline, and diesel emits small amounts of CO_2 at an immense number of places. It is difficult and expensive to capture and store this CO_2. It is much easier and more sensible to provide an energy carrier that produces no CO_2 when it is used, even if the production of such a clean energy carrier entails the emission of CO_2. Capturing CO_2 at a large-scale production plant is much easier than from a multitude of small sources, such as car exhaust pipes.

There are two important energy carriers that produce no CO_2 when used: electricity and hydrogen. A little bit of heat is released as a waste product when electricity is used. The use of hydrogen generates electricity and/or heat, and the "waste product" consists of clean water—which can be reused, certainly in areas with water shortages.

There is yet another energy carrier whose end-use involves no CO_2, namely, hot water or steam. This energy carrier is however only used locally.

We use electricity principally for power and light. We can also use it for transport (e.g., battery-electric vehicles, vessels, and airplanes) and for heat (with a heat pump). We currently use hydrogen as a feedstock in the chemical industry and in refineries. We can also use hydrogen as a feedstock in making synthetic fuels, and to make iron from iron ore. But hydrogen can also be used for mobility (fuel-cell-electric vehicles, vessels, and airplanes), to produce heat with boilers and/or fuel cells, or for the production of electricity with fuel cells when there is a shortage of sustainable electricity.

There are two important energy carriers that produce no CO_2 when used: electricity and hydrogen

An area- and system-oriented energy transition is easier and faster than a sector- and technology-oriented energy transition

Strengthening and repurposing infrastructure

Emission-free end-use requires an extensive infrastructure for the transport and distribution of electricity and hydrogen. Increased use of electricity—for mobility and heating—entails an extension of the existing electricity infrastructure. And for the transport and distribution of hydrogen, apart from the construction of a new hydrogen infrastructure, we also need as quickly as possible to repurpose the existing natural gas infrastructure for hydrogen. The switch from natural gas to hydrogen does not have to happen all at once for the entire natural gas infrastructure. It can also be done in phases, in a process in which you would obviously start by connecting the large-scale hydrogen production sites. Then, one area at a time, you would start repurposing the natural gas infrastructure for the transport and distribution of hydrogen to end-users. To begin with, you would connect areas with large users, for example, industrial clusters. Afterwards, you would let all users in both rural and urban areas switch over from natural gas to hydrogen. Hydrogen will thus flow through the natural gas infrastructure to fueling stations, logistics centers, brick and glass factories, milk and sugar beet factories, bakeries, carwashes, clothing laundries, crematoria, hospitals, schools, homes, etc. And you could then also connect the small-scale hydrogen production sites in these areas to the hydrogen infrastructure.

The natural gas transport companies have developed a hydrogen infrastructure plan for Europe based on the existing natural gas infrastructure. In this plan 75% of the pipelines are repurposed existing natural gas pipelines, and the remaining 25% are to be newly-built hydrogen pipelines [70]. In areas with a less developed natural gas infrastructure the percentage of new pipelines is higher.

Area-oriented transition

Sustainability proposals are often premised on an approach by sector, such as industry, mobility, or homes and buildings. The approach therefore consists of identifying the most appropriate end-conversion technology for the sector concerned. Accordingly, the proposal now is to switch to hydrogen in industry (as a feedstock and for very high-temperature heating), and in heavy transport (ships, airplanes, trucks). For lighter mobility, heating of buildings and homes, and medium temperature for industry, the proposal is to switch over to electricity.

We need to quickly strengthen the electricity infrastructure and repurpose the natural gas infrastructure for hydrogen

This sector-oriented energy transition disregards the fact that the energy system is not organized by sector. In an energy system, it is the energy infrastructure, which provides transport, storage, and distribution, that is the connecting link between the energy production sites and the end-users. This infrastructure is organized on an area or regional basis. And since the energy carriers of the future will for the most part be the same—i.e. electricity and hydrogen—for all sectors, the degree of the interconnection of the different sectors in such an area will be strengthened. An area-oriented approach, following the spatial pattern of the energy infrastructure, is therefore the easiest and fastest route to increasing the sustainability of energy use.

Difference between countries

Depending on the initial situation, an area-oriented approach can look very different from one country to another. Here we compare Norway, where there is a strong electricity infrastructure, with the Netherlands, which has a widely branched natural gas infrastructure. We present this comparison for illustration and in broad outline,

Installation of low-temperature heating

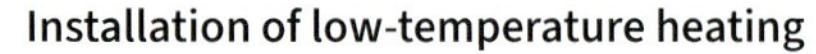

Enhancing sustainability in Norway, with its strong electricity infrastructure, is done differently than in the Netherlands, which has a widely branched natural gas infrastructure

primarily to indicate how a country's context can be taken into account when making choices in the energy system. In Norway, electricity production is 95% based on adjustable hydro power. The existing electricity infrastructure supplies not only power and light, but also space heating. The capacity of the electric grid and the connection capacity at homes are high. This facilitates the switch from electrical resistance heating (which is now extensively used) to heat pumps, since the electric grid requires less capacity for heat pumps than for resistance heating.

The electrification of other functions, such as mobility, is also easy in Norway. The proportion of battery-electric vehicles is high and the charging infrastructure is good. The charging can be done relatively quickly at homes, because these often have a high-capacity connection. You can also easily connect heavy excavators or pile drivers electrically by cable to this strong electric grid.

In the Netherlands the situation is different. There the electricity production is based on gas- and coal-fired power plants, and is being made more sustainable with solar- and wind-generated electricity. The electricity infrastructure is used for power and light, but not for space heating. Gas accounts for more than 95% of the heating. The average capacity of the gas connections used for this purpose is 10 times bigger than that of electricity. Moreover, the existing housing is often poorly insulated. It is therefore not so easy to make existing housing in the Netherlands completely sustainable using electric heat pumps. The electricity infrastructure does not yet have the capacity required for this. When it comes to heating, the switch from natural gas to hydrogen would in principle be easier and faster. The hydrogen would not need to be produced in the Netherlands itself, but could be imported from places with lots of sun and wind. The gas infrastructure is already in place, and difficult-to-insulate homes can quite easily switch over to sustainable sources. Hybrid heating would be even smarter: a small heat pump for baseload heat, for which you wouldn't need to strengthen the electric grid, and a hydrogen boiler for peak loads (winter cold and hot water production)[142].

In the Netherlands hybrid heating with electricity and hydrogen is a smart way of enhancing sustainability, without having to strengthen the electric grid

In the somewhat more distant future, you could produce electricity in buildings and homes with a small fuel cell whenever the solar panels aren't producing electricity—at night and in the winter. You could then use the heat produced by the fuel cell for space heating and hot water production.

You could in the same way increase the sustainability of mobility. Today, there are cars and trucks in the Netherlands that run on natural gas, and the fueling stations are connected to the natural gas network. The natural gas is compressed, so that you can fill up with compressed natural gas (CNG). In the future, hydrogen could be transported through this gas network to the fueling stations. You could therefore fill your tank with hydrogen relatively easily and cheaply. And that would also mean that it would be easier to make construction activities in the Netherlands more sustainable, by equipping trucks, generators, heavy excavators, and pile drivers with a fuel cell. These machines could then simply be filled up with hydrogen at a fueling station.

Norway and the Netherlands are two extreme cases, while many other European countries are situated in between. You have to examine which sustainability approach can be realized quickly and affordably for each area. Also, outside Europe, in places with different climate conditions and societal factors, and usually more limited energy infrastructures, you will always need to reconsider how best to proceed to increase sustainability. But it will be clear that an area- and system-oriented approach will be easier and faster than one based on sectors and technology.

Flexibility thanks to hydrogen storage

The natural gas system is several orders of magnitude larger than the electricity system. Currently, the natural gas system, along with coal and oil, "feeds" the electricity system. Natural gas is transported by pipeline from a gas field to a gas-fired power plant, which converts it into electricity. Thanks to the storage of natural gas in empty gas fields, salt caverns, and porous rock formations, the natural gas system, and thus the electricity system as well, is flexible: the supply can be quickly adjusted to a variable demand. Large-scale natural gas storage, on an hourly, daily, weekly, and seasonal basis, is cheap, and can balance supply and demand on all timescales, for both natural gas and electricity.

Since you can store hydrogen in the same large-scale and cheap manner, it will provide for the weekly and seasonal flexibility in the electricity system. Supercapacitors, batteries, electric boilers, and other forms of demand flexibilization are needed to provide flexibility on the shorter timescales of seconds, hours, and days.

The complete conversion from natural gas to hydrogen means that gas-fired power plants will then run on hydrogen. The plants will therefore no longer emit any CO_2. Still further into the future, more fuel cells will be brought

Features of the current natural-gas and electricity system [4]

	Gas system	Electricity system
Production volume by site	Gas field 10-1,000 TWh/year	Electric power plant 1-30 TWh/year
Distance between production site and use	Pipeline up to 5,000 km Worldwide transport of LNG by ship	Cable Up to 1,000 km
Transport capacity	Pipeline 10-35 GW	Cable 1 (HVAC) - 4 GW (HVDC)
Storage capacity	Salt cavern 200-500 GWh Empty gas field, with 10-100 times greater storage capacity than salt caverns	Pumped hydro power storage 5-25 GWh Largest battery storage system (2023) < 1 GWh

into the system: stationary fuel cells in homes, schools, hospitals, farms, greenhouses, and businesses, and mobile fuel cells in ships, trains, trucks, busses, cranes, tractors, taxis, vans, and cars. All these fuel cells will together provide the electricity system with so much flexibility that, ultimately, large-scale, hydrogen-fired power plants will no longer be needed.

Hydrogen and electricity price competition

It is of course important to quickly produce lots of sustainable energy from all available sources. But from a system perspective, the production of sustainable electricity makes economic sense when it is done relatively “nearby” the end-users, since the transport and storage costs of electricity are high relative to its production costs. With hydrogen things are different. Sustainable hydrogen production makes economic sense when it is done on a large scale, in places with good sustainable energy sources—primarily sun and wind—as well as lots of cheap space. But these places are often “far away” from the end-users, in deserts or on the sea or ocean. In the case of hydrogen, the transport and storage costs are actually low relative to the production costs. For hydrogen, lower production costs are therefore of greater significance for the price than lower transport and storage costs.

When we say “nearby” we mean a few hundred up to, at most, a thousand kilometers from the demand, in a place where you can produce sustainable electricity in the order of kilowatts to gigawatts. The transport distance is limited by the transport costs, the scale of the electricity production by the capacity of the cables. For all forms of sustainable electricity production, you will consider whether the costs of electricity production and electricity transport can in fact compete with the cost of producing electricity with imported hydrogen.

Stationary and mobile fuel cells provide the electricity system with so much flexibility that hydrogen-fired power plants will ultimately no longer be needed

By “far away” we mean distances from hundreds to thousands of kilometers from the demand. Hydrogen transport by pipeline can be economically attractive for distances up to about 5,000 kilometers. For distances up to a few thousand kilometers, transport by ship is more costly than by pipeline, mostly because an extra conversion from gas to liquid is still required. What is important is that the hydrogen transport takes place as continuously as possible, in baseload. This means that storage possibilities also need to exist in the vicinity (up to a distance of a few hundred kilometers) of the hydrogen production site—for example, in the underground, in salt caverns, empty gas fields, rock formations, or aquifers. An alternative is to store hydrogen above-ground in tanks, in liquefied or compressed form, or as ammonia. Storage above-ground is significantly more expensive than storage underground.

In the case of large-scale hydrogen transport by pipeline, we are talking about several million tonnes of hydrogen annually. To load pipelines with a transport capacity of 10-30 GW, several tens of wind GW to hundreds of solar GW are needed. This refers to the export of hydrogen from areas where cheap hydrogen can be produced; and to the import to areas where energy demand is large, and where it can’t be produced, or can’t be produced at a competitive price.

The advantage of transporting hydrogen over long distances by ship is that this offers supply flexibility. Thus, in the case of hydrogen production using floating wind and solar farms far away on the oceans, we will see that the hydrogen will be transported by ship (possibly in other forms, such as ammonia). These vessels will sail to the location where the best prices are being paid for hydrogen. This resembles for instance the way today’s LNG market works. A hydrogen world market will develop, like the one we have for oil and gas, with regional markets for electricity. Price competition will determine which energy carriers, in what proportions, are used in a specific area and sector.

A hydrogen world market will develop; for electricity regional markets

A large gas transport pipeline through the desert

Small-scale hydrogen production “nearby” can also be useful. If the electric grid is congested, you can still generate more sustainable energy, by producing hydrogen and delivering it to the hydrogen infrastructure. In this manner you can cheaply transport and store hydrogen for its later use when there is a demand for it. And at particular sites, such as wastewater treatment plants, you can also make good use not only of the hydrogen, but also of the generated oxygen and residual heat. Moreover, in specific situations, it is economically attractive to produce hydrogen close to the demand, because this allows you to save on hydrogen transport costs by tube trailer—for example, for hydrogen fueling stations that are not connected to a hydrogen pipeline. In all of these situations the hydrogen production is directly linked to a sustainable energy source, and the scale of the hydrogen production ranges from tens of kW to tens of MW.

Conversion at-source: from natural gas into hydrogen

A rapid switch of the natural gas infrastructure to a hydrogen infrastructure involves two challenges, namely, how to quickly produce enough hydrogen, and what to do with the natural gas. Both challenges can be solved in the same way: by converting the natural gas into hydrogen at-source. Natural gas can be converted into hydrogen using a variety of technologies. The major share of hydrogen produced today comes from steam methane reforming. This process releases CO_2, of which you can capture and store 50-80%. With the newer autothermal reforming technology, you can capture and store 90 to almost 100% of the CO_2. But there is also a technology that emits zero CO_2: methane pyrolysis. In this case, methane is heated and separated into hydrogen (H_2) and solid carbon (C) [143]. Carbon is a raw material for which there is a strong demand: it is used in the production of ink, paint, graphene, and car tires, as well as being a soil conditioner. Besides, you can store it easily and cheaply.

Powdered carbon

Converting natural gas into hydrogen at-source has several benefits. For instance, the CO_2 emissions can be immediately captured and stored in the gas field, with no need for an extensive CO_2 infrastructure. Better still, is to produce absolutely no CO_2 by using methane pyrolysis. And since extensive technical-chemical installations already exist at the gas fields—among others, to clean the gas, mix it with nitrogen, and compress it—you can also easily produce hydrogen there. Furthermore, you can build a large installation with scale benefits that makes it cheaper than small installations. In this way you also avoid the methane emissions in the chain of transport, storage, distribution, and use.

Price competition will determine how, where, when, and how much hydrogen and electricity is used

The conversion of natural gas into hydrogen is also called pre-combustion CCS (carbon capture and storage). This process occurs before the natural gas combusts. By using steam to convert natural gas into hydrogen and CO_2, you get a pure CO_2 stream that can be effectively captured. CCS is often thought of in terms of capturing CO_2 from the flue gases of electric power plants, large steam boilers, blast furnaces, or chemical installations. This is called post-combustion CCS: the process takes place after the "combustion" of natural gas, crude oil, or coal. You first need to separate the CO_2 from the waste gases—mixtures of CO_2, nitrogen oxides, and other gases—and then transport and store it. This often concerns smaller installations, while you need a large CO_2 infrastructure for the transport and storage. Pre-combustion CCS at-source is therefore preferred over post-combustion CCS.

Methane pyrolysis

Hydrogen production with zero CO_2 emissions

We currently make hydrogen from natural gas or biogas (methane), which involves the emission of lots of CO_2. But there is also a hydrogen production process that emits zero CO_2: methane pyrolysis. In this process, methane is heated to high temperatures in the absence of air (oxygen). At a temperature of more than 1,200 °C, the methane splits into hydrogen (H_2) and solid carbon (C) [143]. These are easy to separate because hydrogen is a gas and carbon a solid substance. If, for the heating process, we use green electricity or some of the produced hydrogen, then absolutely no CO_2 is emitted in the production of hydrogen with methane pyrolysis.
In theory, the splitting of methane requires seven times less energy than the splitting of water. However, methane already has a certain economic value, which means that the total energy balance of hydrogen made from methane is less positive than that of hydrogen made from water.

Monolith, methane pyrolysis plant in Nebraska, USA [145]

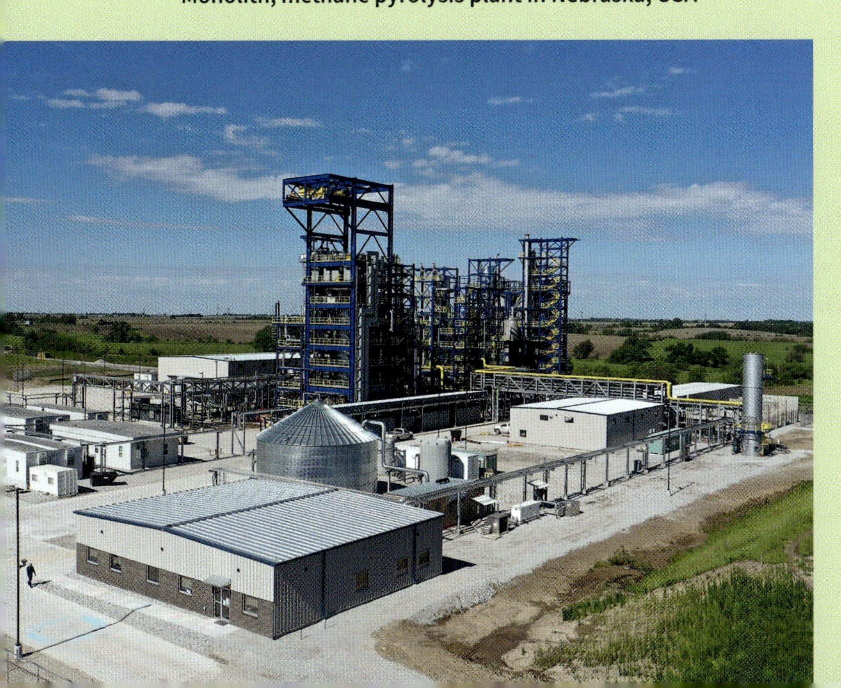

Various companies, research institutions, and universities worldwide are developing methane pyrolysis technologies. One outcome of this effort is the reduction of the splitting temperature through the use of catalysts. But this is often at the expense of the purity of the carbon [144], while it is pure carbon that is needed as a raw material for ink, paint, graphene, nanotubes, strong carbon fibers, supercapacitors, and car tires. Moreover, pure carbon is a good soil conditioner, which better retains water and nutrients in the soil.

The methane pyrolysis of natural gas produces hydrogen and solid carbon, so no carbon dioxide is emitted

The Monolith company has an operational methane pyrolysis plant in the US state of Nebraska. The methane is split into hydrogen and carbon in a plasma arc at a temperature of 2,000 °C. The plant was actually built to produce pure carbon for car tires. For this reason, the energy efficiency of this hydrogen production process is still quite low.

Conversion of biogas into H_2 and CO_2

Energy from biomass residues can constitute an important source of sustainable energy in certain areas. Biogas can be directly used for the production of heat and electricity in boilers or gas engines. Biogas is also increasingly being upgraded to natural gas quality or to biomethane, after which it is fed into the natural gas infrastructure or transported as bio-LNG for use in combustion engines. Just like we converted natural gas at-source into hydrogen and CO_2, we can also do this with biogas, which moreover already contains a percentage of CO_2. This results in green hydrogen and green CO_2. You can use both hydrogen and CO_2 as feedstock; and hydrogen additionally as an energy carrier. You can feed hydrogen into a hydrogen infrastructure, or transport it by road with tube trailers (trucks that transport high-pressure cylinders filled with hydrogen), or cryogenic transport trailers (trucks that transport liquid hydrogen). You can also liquefy CO_2 and transport it by road in tanker trucks. Both CO_2 and hydrogen are tradeable products. Because of the competition with solar and wind hydrogen, the price of this hydrogen will probably decrease, but the price of CO_2 is actually expected to increase. On balance, you therefore end up with a positive business case for biogas. Also, through biogas conversion into hydrogen and CO_2, you avoid a gas lock-in, a situation in which you need to preserve a gas infrastructure for a limited volume and for a longer period of time.

Government direction for rapid conversion

Because energy is one of the basic necessities of life, governments have always exerted a strong influence on the energy provision. This changed in many western countries at the end of the last century, when the energy markets were liberalized to a larger or lesser degree. Since then, energy prices have been determined by the market, where supply and demand are matched. Can a rapid transition from a fossil to a sustainable energy system be realized by adjusting the market mechanisms, for example, by creating a market for CO_2 emissions? This is perhaps a possibility, yet it probably won't lead to sufficiently rapid change. Concretely, more is needed than a gradual, incremental adjustment of the energy system. What is needed is a fundamental change. The conversion to a sustainable energy system requires that the supply, the demand, the infrastructure, and the storage of energy all be developed simultaneously and conjointly, starting practically from zero. This calls for government direction.

Pre-combustion CCS at-source, in which hydrogen is produced, is preferred over post-combustion CCS

Governments also played a leading role in the development of the electricity and gas system in the past. The same applies to the conversion from coal to gas, and from town gas to natural gas. Mature markets for sustainable energy will only be able to develop once the system disposes of sufficient infrastructure and storage capacity, and there is sufficient supply and demand.

A rapid transition from a fossil to a sustainable energy system does not have to cost the society more than a slow conversion. Moreover, it leads to much lower greenhouse gas emissions. In a large-scale, rapid conversion, the costs of sustainable energy production, transport, storage, and end-conversion technology will drop faster. The drop in the costs of technology is described in learning curves, which show the percentage drop in the cost price for every doubling of the sales of the equipment—for energy, this concerns the total installed capacity, for instance of solar panels, electrolyzers, or heat pumps. The faster you convert, the faster the prices will drop. And since you emit less greenhouse gases, you pay less total carbon taxes [146].

The transition from a fossil to a sustainable energy system is a fundamental and urgent change. A change that will not happen on its own. That is why it is necessary that governments direct this transition and, when necessary, even organize it, in collaboration with companies and citizens, of course.

Hyundai's self-driving hydrogen truck

Through biogas conversion into H_2 and CO_2 a natural gas lock-in is avoided

Green energy for all

Based on the current socioeconomic and technological developments underlying the IPCC's Middle of the Road scenario, we outline here a global picture of a sustainable energy system in 2100, taking account of the time and space dimensions. We also outline what the total sustainable system for energy, materials, water, and food could look like.

Socioeconomic scenarios for 2100

The Intergovernmental Panel on Climate Change (IPCC) works with several future scenarios or Shared Socio-economic Pathways (SSPs) for its climate models. Each scenario applies different assumptions regarding global population growth, access to education, urbanization, economic growth, resources availability, technology developments, and drivers of demand for products and services, such as lifestyle changes [147]. Depending on which SSP you use, the modelled global warming in 2100 ranges from 3.0 °C (SSP1) to 5.1 °C (SSP5) above the pre-industrial level.

Key elements of the Shared Socioeconomic Pathways (SSPs) in 2100 [148]

	SSP 1 Sustainability	SSP 2 Middle of the Road	SSP 3 Regional rivalry	SSP 4 Inequality	SSP 5 Fossil fuel development
Global population (billion) [149]	7.0	9.0	12.6	9.3	7.4
Global GDP (billion US$2005/ year) [150], [151], [152]	565,390	539,332	270,265	352,091	1,031,000
Urbanization (%) [153]	92.6%	79.7%	58.4%	91.7%	93.0%
Primary energy use (EJ/year) [154]	701	1,304	1,215	903	1,824
Greenhouse gas emissions (Mt CO_2/year)	24,613	85,684	85,215	44,785	126,098
Temperature increase (°C)	3.0	3.8	4.1	3.8	5.1

On the basis of these SSPs, the IPCC developed scenarios with policies to meet the goals of the Paris Agreement. In 2015, it was agreed in Paris that the global average temperature increase would be kept well below 2 °C relative to the pre-industrial level [155].

Energy use per km² in 2100

To begin with we mapped out the energy use per km² for the year 2100, using SSP2 as our premise (i.e., with no additional climate policy). Using figures on population growth, urbanization, and primary and final energy use, we calculated the amount of energy use per capita for each of the continents.

Carbon needs in 2100

Fossil fuels are not only used as a source of energy, but also as feedstock for chemical products and synthetic fuels. The crude oil, natural gas, and coal fossil fuels consist of carbon and hydrogen atoms. With these atoms you can, together with oxygen and nitrogen from the air, produce all chemical products, although you need to add a variety of other elements in much smaller quantities.

Most of the carbon locked in chemical products is released at the end of their life span in the form of CO_2. In the case of synthetic fuels this happens immediately during the combustion. This is of course not desirable. But if we generate energy with solar, wind, and hydro power, another problem arises: a shortage of carbon. This is why we need in the future to find other, renewable, and sustainable sources of carbon, for example, in recycling, biomass, or biomass residues, and possibly even from the air.
In our daily lives we use many products from the chemical and associated industries.

Energy use per capita in 2100 according to SSP2 [156]

	World	Asia (excluding Japan and the Middle East)	Latin America and the Caribbean	Middle East and Africa	OECD-countries (OECD = Organisation for Economic Co-operation and Development)	Eastern Europe and Russia
Population (billion)	9.0	3.76	0.67	3.09	1.27	0.24
Urbanization (%)	79.7%	75.8%	93.2%	75.2%	93.9%	87.3%
Primary energy use (EJ/year)	1,293	474	118	359	283	59
Final energy use (EJ/year)	962	361	76	279	209	37
Primary energy use (MWh per capita, per year)	39.7	35.0	49.0	32.3	61.8	67.4

Products from the chemical industry [157]

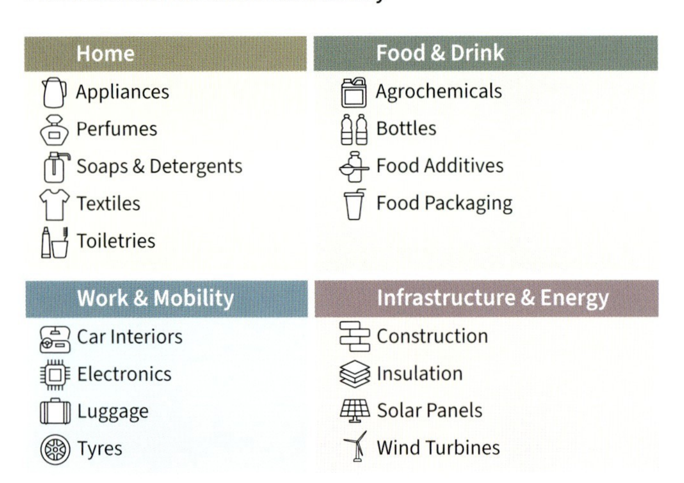

In 2020 the demand for carbon for chemical products was about 450 million tonnes. Approximately 85% of this carbon came from fossil fuels, 10% from biomass, and only 5% from recycling [15]. The Nova Institute estimates that the global demand for carbon in 2050 will be 1,000 million tonnes, of which 55% will come from recycling, 20% from biomass, and 25% from CO_2 that is captured from industrial processes and the air [157]. In the scenario for the year 2100 that we are using here, we also assume a demand of 1,000 million tonnes of carbon, of which 55% will come from recycling, and 45% from biomass residues and seaweed. The 1,000 million tonnes represents 3,700 million tonnes of carbon dioxide.

Carbon is also needed to make synthetic fuels for aviation and shipping. Small aircrafts and vessels will possibly be powered electrically, and medium-sized to large ones will fly and sail with liquid hydrogen. Large aircraft and vessels that cover long distances can't however carry enough energy in the form of liquid hydrogen. Aviation will need synthetic aviation fuel, also known as sustainable aviation fuel (SAF), while shipping will use synthetic diesel or perhaps methanol. To make these fuels, about 2,000 million tonnes of carbon dioxide will be needed in 2100, that is, 550 million tonnes of carbon. The carbon for synthetic fuels can't be derived from recycling, so that it will come from biomass residues or seaweed, so-called short-cycle carbon. The CO_2 emitted in the combustion of these fuels on-board can't be effectively captured and stored onsite.

We already noted earlier that it is economically attractive to fully convert biomass residues and seaweed into hydrogen (H_2) and carbon (C), or carbon dioxide (CO_2). Cultivating seaweed and converting it into hydrogen and CO_2 is probably less costly than the direct air capture of CO_2.

The same applies to both wind and solar: shortages in densely populated areas, surpluses in sparsely populated areas, and little to no energy production in agricultural, nature, mountainous, and forest areas

Supply and demand of carbon for chemical materials, products, and synthetic fuels in 2100 [157]

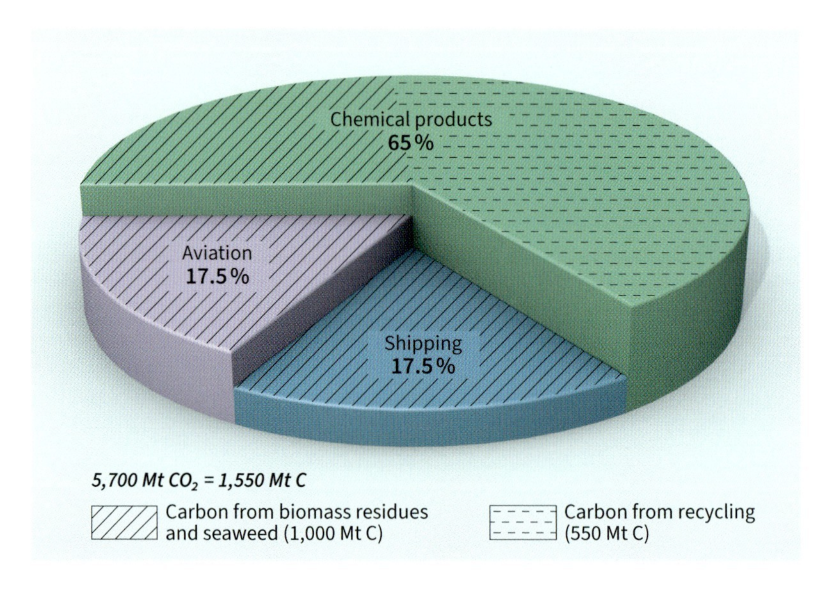

Solar energy heat map: surpluses and shortages per km² in 2100 [156], [40], [41], [42], [43], [158]

Tropic of Cancer

Equator

Tropic of Capricorn

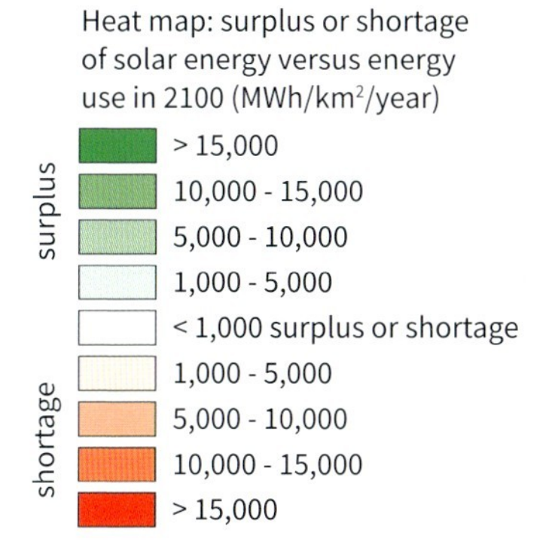

Surpluses and shortages of sustainable energy

As we did for 2020, we have also made a solar energy heat map for 2100. The map shows that there is not only a solar energy shortage in Europe, on the east and west coasts of the United States, and the east coast of China, but also in areas with rapid increases in population and/or standard of living. This concerns Asian countries such as India, Pakistan, Thailand, Indonesia, and the Philippines, but also African ones, including Nigeria, Ethiopia, and Uganda, and Central-American countries like Mexico and Guatemala, and South-American ones like Colombia. Central Africa faces big shortages in 2100, even though it receives lots of solar irradiation. This is because of the presence of tropical rainforests, which we of course won't cut down to install large-scale solar farms. The limited space for solar farms, along with the population growth, explain the shortage of solar energy.

The heat map shows where the surpluses and shortages are, but says nothing about the level of the production costs. Even though there are solar energy surpluses in countries like Canada and Russia, it is not likely that large-scale solar farms will be built there. The intensity of the solar irradiation in these countries is after all low, which means the production costs per kWh are high. We haven't made a heat map for wind energy, but it would look similar to the solar energy heat map. For wind energy, even more so than that of solar energy, it is the case that none, or very little, can be generated in densely populated areas. But you can, to a certain degree, combine wind energy with agricultural production, which is not the case for solar energy. In general, however, the same applies to both wind and solar: shortages in densely populated areas, surpluses in sparsely populated areas, and little to no production in agricultural, nature, mountainous, and forest areas.

The oceans are not on the world solar heat map for 2100, but they are certainly important for the production of solar and wind energy. It is also true for the oceans that not all areas are available for energy production. Think for instance of nature areas, like the Sargasso Sea, fishery areas, shipping routes, and anchorages near the coast. But, all things considered, a world renewable energy heat map will show a surplus on the oceans. Solar and wind production costs on the oceans, as on land, are not the same everywhere. You may expect the lowest costs in areas with high wind speeds and/or solar irradiation, within a few thousand kilometers from the coast. The surplus in such areas is large, and the production plus transport costs are economically attractive.

Large-scale offshore wind farm

Final energy and carbon use in 2100

Year 2100	World	Electricity	Hydrogen for energy and feedstock	Heat
Final energy use	100%	50%	40%	10%
	974 EJ	487 EJ	390 EJ	97 EJ
	= 270,500 TWh	= 135,250 TWh	= 108,200 TWh	= 27,050 TWh
Year 2100	**World**	**Chemical products**	**Synthetic aviation fuel**	**Synthetic shipping fuel**
Carbon use	100%	65%	17.5%	17.5%
	1,550 Mt C	1,000 Mt C	275 Mt C	275 Mt C
	= 5,700 Mt CO_2	= 3,700 Mt CO_2	= 1,000 Mt CO_2	= 1,000 Mt CO_2

Electricity and hydrogen as energy carriers

What we learn from this heat map is that not enough sustainable energy production can or will occur in densely populated areas, thus in cities. So, to provide cities with energy, the sustainable energy has to be transported to them. Sustainable energy production, in the form of electricity, in a radius of hundreds of kilometers around cities is probably not sufficient nor cheap enough. That is the distance which you can bridge economically with an electricity cable. At greater distances, in the deserts and on the oceans, there is indeed sufficient sustainable energy available, but you will have to transport it to urban areas in the form of hydrogen, or hydrogen-derived products. Whatever a sustainable energy system ends up looking like, a lot of energy will be transported, regionally, continentally, and globally. And even if the final energy use consists for the most part of electricity, and cities are fully electrified, the transport and storage of energy will mainly occur in the form of hydrogen. The transport and storage of hydrogen and electricity is cheaper, more reliable, and with higher security of supply than only one energy carrier: electricity.

Energy, carbon, and drinking water balance

We have formulated a global balance for energy and carbon in 2100, bearing in mind the spatial distribution of energy surpluses and shortages (from the world solar heat map) and the variations of supply and demand through the year. Of course, this initial rough estimate can be further improved and detailed. We only use it here to illustrate the influences of the dimensions of space (transport of energy) and time (storage of energy) in a sustainable energy and carbon system.

According to the Middle of the Road scenario (SSP2), global final energy use in 2100 will be 270,500 TWh. We have divided this amount into the sustainable energy carriers electricity (50%), hydrogen (40%), and heat (10%). As a complement to the Nova Institute's estimate for carbon use for chemical products (1,000 Mtonnes or Mt), we have made a carbon-use estimate of our own for synthetic fuels for aviation (275 Mt) and shipping (275 Mt). Altogether, the global carbon use in 2100 will then be 1,550 Mt.

Influence of the space dimension

The world solar heat map shows that sustainable energy needs to be transported over short and long distances, from areas with surpluses to areas with shortages. Each energy carrier has different transport costs. Because heat transport in the form of water is the most expensive, you want to generate and use heat locally. Electricity transport is less costly, so that you can generate and use regionally. The transport is done through electricity cables. Even less costly is the transport of hydrogen in the form of gas via pipelines, so that it is cost-effective to transport it over thousands of kilometers (thus also globally). If hydrogen gas is made into a liquid, it can be transported worldwide by ship. Apart from liquefied hydrogen, you can also transport liquid or solid hydrogen-derived products, such as synthetic fuels, ammonia, or hot briquetted iron.

Seaweed and biomass residues contribute not only to the energy balance, but also to the carbon balance. In a sustainable energy system, carbon no longer comes from fossil sources, but from biomass or recycling. That's why we want to convert the available biomass into hydrogen and CO_2. We bring the seaweed that is cultivated at sea to shore, where we use it to make hydrogen and CO_2 and/or syngas (a gas mixture of carbon monoxide and hydrogen gas). We then process these into chemical products and fuels for ships and aircraft. We make use of wet and dry biomass residues on a regional basis for the production of hydrogen and CO_2 or C (carbon). With dry biomass residues we can also produce heat locally, by combusting them in stoves, ovens, or fireplaces.
You can produce low- and medium-temperature heat with electricity and/or hydrogen. But you can also extract this heat directly from geothermal sources, solar energy, and biomass. And you can also use residual heat. In the future, one of the key local sources of residual heat will be fuel cells. Along with batteries, fuel cells in homes and buildings provide for flexibility in the electricity system. At the same time, they produce heat which you can use for space heating and hot water, or—at medium temperature—for clothing laundries, carwashes, etc.

Influence of the time dimension

Energy storage in a global sustainable energy system has several general features. First of all, the storage must be located as closely as possible to the large-scale production locations of sustainable hydrogen. This enables the transformation of the variable hydrogen supply into a continuous supply, which can then be processed and transported by pipeline or ship. Secondly, hydrogen storage is needed near the energy demand, which enables a response to fluctuating demand due to day-night variations, seasonal variations, and variations caused by weather, living, and working patterns. Hydrogen storage will most of all be able to absorb the large-scale seasonal, monthly, and weekly variations. Thirdly, storage is needed to balance the regional electricity system. The regionally generated electricity supply, which is subject to time variations, can thus be aligned to the electricity demand. This requires storage on all timescales. For variations ranging from seconds to a week, this can be done with batteries and supercapacitors—capacitors that can charge and discharge quickly, and can store 10 to 100 times more energy than electrolytic capacitors. Weekly to seasonal variations require large-scale storage, such as pumped hydro power and hydrogen. Lastly, storage of heat and various fuels is needed at a local level. Heat storage for space heating is needed on various timescales, including seasonal storage in aquifers.

Storage has been the object of numerous studies, but they are all fragmented and focus solely on one part of the energy system. No research has yet been conducted on storage in a global, fully sustainable energy system. This is why our estimates here of the global storage volume requirements are based on a few fragmented studies: on how you can convert the variable supply of solar and wind energy into a continuous supply with hydrogen [69], [159]; on the current fossil storage capacity, for natural gas, among others [160]; and on large-scale seasonal storage in fully sustainable electricity systems [161], [162]. On p. 143, we summarize how much of the hydrogen and electricity produced needs to be delivered by, or via, a storage system. These are very rough estimates.

Amount of energy needed to be drawn from storage in a global sustainable energy system to satisfy energy demand at all times

Storage objective	Technology	% of production/final demand via storage
Hydrogen production: from variable to continuous or baseload supply	Hydrogen storage, underground and above-ground, near hydrogen production to enable transport of baseload	35%
Hydrogen demand: from baseload supply to variable demand	Hydrogen storage, underground and above-ground, near hydrogen demand	25%
Electricity balancing (hour, day, week)	Electricity storage in supercapacitors and batteries	10%
Electricity balancing (week, month, season)	Electricity and heat production with fuel cells on hydrogen	25%

Energy and carbon balance in 2100

The dimensions of space (regarding transport) and time (regarding storage) play a bigger role in a sustainable energy system than in a fossil energy system. A fossil energy system is primarily about the energy production, divided into the various sources, and the energy demand, divided into various sectors. In a fully sustainable energy system, it is also about the dimension of space, which you can divide into the global, regional, and local categories, as shown in the figure on p. 144. Based on the SSP2 scenario (Middle of the Road), and taking the dimensions of time and space into account, we have calculated a global sustainable energy and carbon balance for the year 2100.

Let's look first of all at the carbon balance in 2100. We assume that 55% of the carbon sequestered in chemical products can be recycled. The other 45% needs to be produced from new, green carbon. The production of synthetic fuels also requires new carbon, because when it is combusted in ships and aircraft the carbon disappears into the air. In total, we will need about 1 billion tonnes of carbon per year from seaweed and biomass residues, which is equivalent to 3.7 billion tonnes of CO_2. We assume that the seaweed produced on the sea will supply half of this carbon. The other half will come from a diverse supply of local biomass residues, such as sludge, agricultural waste, or wood waste. Together with recycled carbon, this will be processed into chemical products and synthetic fuels.

The energy balance in 2100 is a little more complicated. We examine it on three spatial scales:

1 **Global.** Large-scale hydrogen production takes place in deserts and on the oceans (tens to hundreds of GW per location), using primarily sustainable energy sources and possibly nuclear energy. The hydrogen is produced by electrolysis, photolysis, or thermolysis. Hydrogen and carbon production from seaweed also occurs globally, after which these substances are converted into chemical products and synthetic fuels. A continuous supply of hydrogen exists thanks to storage. It is transported by pipeline or ship to areas with an energy demand, such as cities, towns, and industrial locations. At the large-scale hydrogen production locations, sustainable electricity is used, for compression and to convert hydrogen and carbon into other products, among other purposes.

2 **Regional.** On this scale, it is mainly electricity that is generated. Transport and distribution occur through

Worldwide energy and carbon flows in 2100

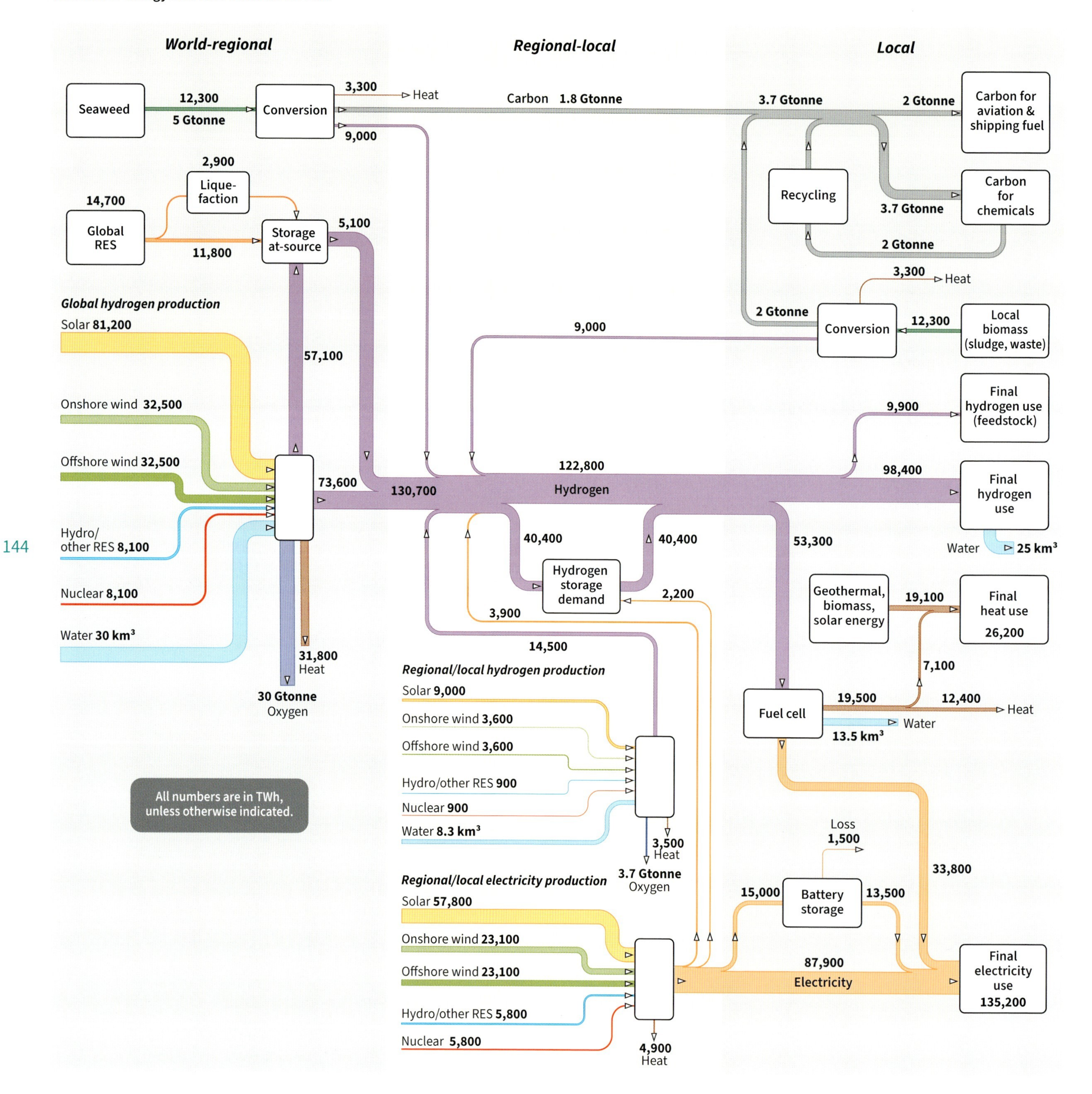

an electric grid. The balancing of the electric grid is accomplished with batteries (short term) and with fuel cells (monthly and seasonal variations). Regionally stored hydrogen, together with hydrogen produced at the instant, is used to make it possible to meet hydrogen demand at all times. Part of this demand comes from stationary fuel cells that balance the electric grid. Hydrogen is also produced on a smaller scale, among others, from biomass residues (which in the process also produce carbon), at water treatment plants (together with oxygen and heat production), and at the locations of hydrogen demand (if the hydrogen can't be delivered through a hydrogen infrastructure).

3. **Local.** On this scale, it is electricity, heat, and hydrogen that are produced. Electricity is produced with solar cells or fuel cells. Hydrogen is produced using reversible fuel cells or photolysis cells, which use sunlight to split water into hydrogen and oxygen. Local production only represents a small part of the total hydrogen supply. Heat is generated mostly on the local scale, because its transport over long distances is not possible without large losses. Heat is produced from sources like solar energy, geothermal energy, and biomass. Fuel cells that work on hydrogen also produce heat.

The final electricity use, representing 60% of final energy use, is intended for the operation of appliances, light, cooling, heating with heat pumps, and battery-electric mobility. The final hydrogen use, 40% of the total, is intended as feedstock for chemical products and synthetic fuels, for high- and medium-temperature heat in industry, and for fuel-cell-electric mobility. Part of the final heat demand in homes, buildings, greenhouses, and industry is supplied directly through heat sources or through the heat from stationary fuel cells. Stationary fuel cells on hydrogen play a crucial role in balancing the electricity system and in supplying heat. Smart combinations of hydrogen and electricity, and sometimes also of heat, result in an affordable, reliable, and clean energy provision with a high security of supply.

Transport of water and energy

In 2100 a large part of the hydrogen is made from water through electrolysis, photolysis, or thermolysis. If this hydrogen is chemically converted in a fuel cell into electricity and heat, demineralized water is released. By capturing this water, and adding salts and minerals to it, clean drinking water is produced which contains no contaminants whatsoever. Stationary fuel cells are always installed locally, close to the demand for electricity and heat. Therefore, drinking water can be produced locally too. You can use this method to produce sufficient water on a cargo ship for the crew, both for consumption (drinking water) and sanitation (toilet flushing). A hydrogen car can also produce substantial quantities of water: 1 kg of hydrogen (enough to drive 100 km) generates 9 liters of very clean water (3 days of drinking water for 1 person). With hydrogen you don't only transport energy, but also clean water!

Hydrogen cars produce 9 liters of very clean water—3 days of drinking water for 1 person—from 1 kg of hydrogen (100 km drive)

How much drinking water is needed globally, and how much of this can stationary fuel cells produce, along with electricity and heat? If we assume an average consumer use (for cooking and drinking) of 3 liters a day, and that the population in 2100 is 9 billion, then nearly 10 billion m^3 of drinking water is needed. According to our energy balance, this level of drinking water consumption is below what the locally installed, stationary fuel cells produce: about 13.5 billion m^3 of water. Especially in areas with water shortages, fuel cells can in the future be a welcome source of drinking water for consumption.
Drinking water consumption represents only a fraction of total water use. In Europe we daily use about 150 liters per person for showering, toilet flushing, and washing[118].

The assessment of a sustainable energy system requires a balance of materials and of energy, with the primary materials use as input, and the final energy use as output

This demand could drop through the use of water-saving appliances (shower, toilet, washing machines, etc.). And plenty of energy—over 60%—can be recovered from shower water [163]. Also, there are many local possibilities for drinking water reuse, and we can—besides tapwater—use rainwater as a source [164]. Seawater, transported over long distances, is also a possible source of freshwater. By far most freshwater, 70%, is used for the irrigation of agricultural crops [165]. But, as the World Bank argues, we will have to start growing much more food on the sea. Food from seaweed needs no irrigation water.

Sustainable energy system goals in 2100

In our outline of a sustainable energy system in 2100, we have paid attention to a few sustainable energy system goals: clean (no greenhouse gas emissions), affordable (lowest system costs for production, transport, and storage), and reliable (storage, system flexibility, and redundancy). In this initial effort, we have incorporated the space and time dimensions as much as possible. But the future energy system must naturally satisfy all sustainable energy system goals. It should therefore also be circular and fair, and have energy- and materials-supply security. These themes have already been the subject of reflection [166], but a good assessment requires even better and more detailed methodologies and instruments.

What is clear in any event is that the future sustainable energy system is not so dependent on the availability of energy sources. Sun and wind are abundant. But the system does rely on the availability of materials needed for the conversion of sustainable energy into energy carriers, for transport and storage, and ultimately for the conversion of energy carriers into energy services. These materials are scarce. This has an impact on two goals: circularity and materials-supply security. Below, we propose the outline and methodology to enable an assessment of these two goals.

Today we usually assess energy systems on the basis of an energy balance. This is what we have done above for a sustainable energy system in 2100, where we also included the space and time dimensions. But for a sustainable energy system, this sort of energy balance is actually misleading.

In an energy balance we compare the input of energy (the primary energy use) with the energy that we use (the final energy use). If you divide the final energy use by the primary energy use, then you have a measure of the energy efficiency. Many see this as an important measure: the higher the efficiency, the better the energy system. The question is whether this also applies to a sustainable energy system. Because, how do you actually determine how much energy we put in?

In a fossil energy system the amount of energy we put in is clear: it is the energy content of the fossil fuel, or the amount of energy used in the form of oil, coal, and gas. After its use, this form of energy no longer exists. With sustainable energy sources, like solar or wind, things are different. Is the amount of energy that we put in the system equal to an amount of used solar energy? That sounds logical, but the solar energy does not run out because of this use. You can also begin with the amount of electricity produced by the solar cells. But solar cells in the desert produce more electricity than those on a roof in Germany. This is therefore not a uniform quantity. Moreover, photolysis cells produce hydrogen and not electricity. And there are efficiency differences between solar cells, so that, to produce a given amount of energy, you sometimes need more and sometimes fewer solar cells.

Green energy for all is no utopia, thanks to hydrogen and electricity as clean energy carriers!

The Airbus ZEROe will fly on hydrogen

For these reasons, you can't in fact unambiguously determine the primary energy use of a sustainable energy system. But what is the alternative? Earlier we noted that the transition from a fossil to a sustainable energy system implies that we switch from fossil energy to materials. Shouldn't we therefore replace primary energy use in the assessment with primary materials use? An initial, simple adjustment of the energy balance would therefore involve replacing the primary energy use with installed capacity and fossil energy use—this is after all a measure of the materials and energy use, with the recycling of materials not yet taken into account. In a sustainable energy system, it is a matter of installed capacity for solar, wind, hydro, and biomass conversion; while in a fossil energy system it is a matter of installed capacity for oil and gas extraction, coal mines, refineries, and electric power plants, plus the fossil energy use. The energy balance therefore becomes a balance of materials and energy, with the primary materials use plus fossil energy use as input, and the final energy use as output. With this, we are effectively in a position to assess whether a sustainable energy system is circular, or whether it is affordable, and the degree of materials-supply security we have.

Artist impression of a sustainable energy system in 2100

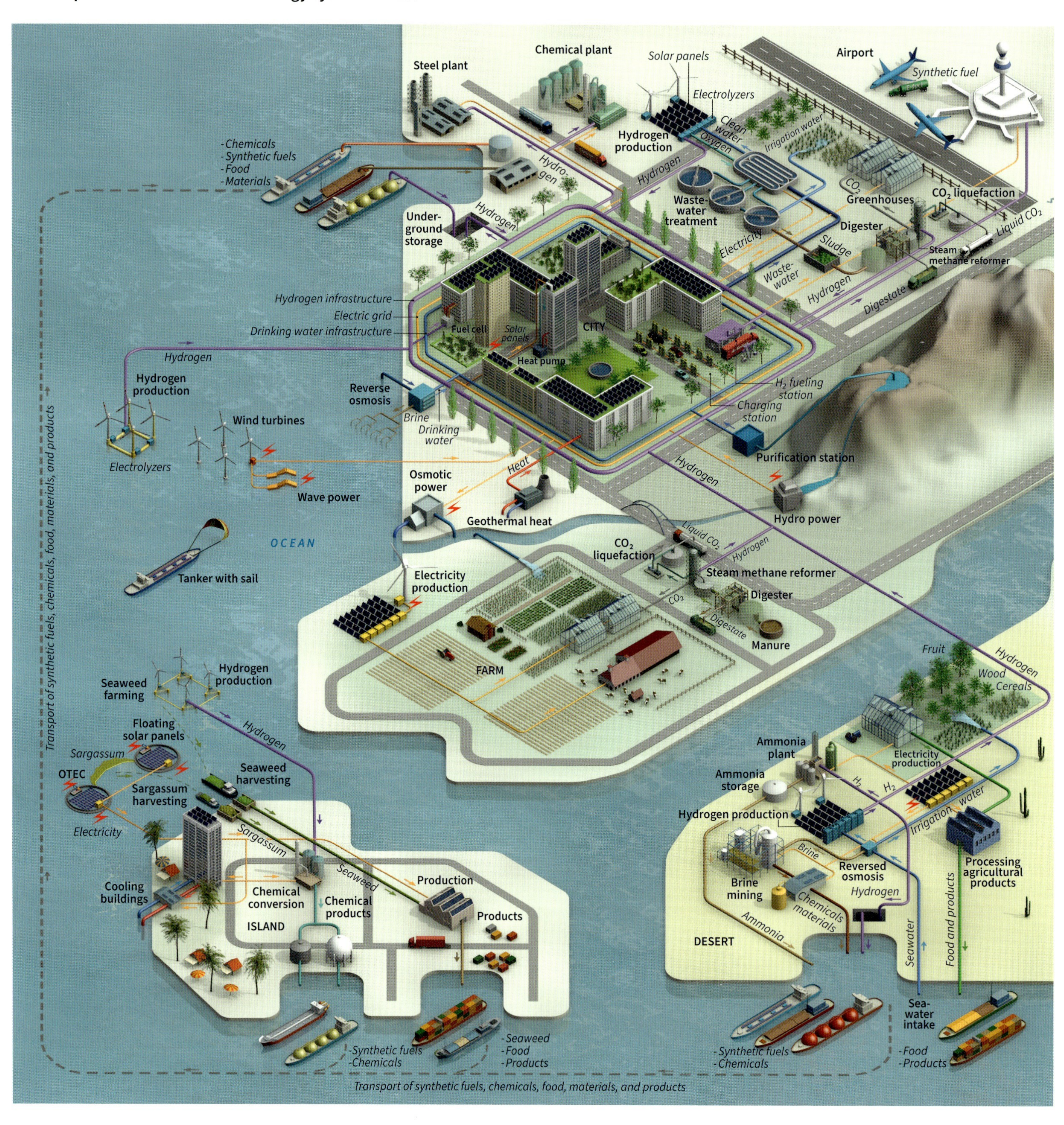

Green energy for all

We have sketched a sustainable energy system for 2100, taking into account the dimensions of space and time. From the world heat maps we concluded that we must import lots of cheap solar and wind energy, in the form of hydrogen, from the deserts and the oceans. And since the supply of sustainable energy fluctuates, a sustainable energy system needs to have more storage than a fossil energy system. The underground offers possibilities for cheap, large-scale, long-term storage. In addition, small-scale, short-term storage is essential for a reliable sustainable energy system. The two important sustainable energy carriers, which produce no CO_2 when used, are hydrogen and electricity. Both energy carriers can moreover be affordably distributed to all end-users: hydrogen through large-scale, long-distance transport, electricity through regional transport. The picture is clear: green energy for all is no utopia, thanks to the clean energy carriers hydrogen and electricity!

References

[1] K. Blok and E. Nieuwlaar, *Introduction to Energy Analysis*, 2nd ed. Routledge, 2017.

[2] P. M. van Oirsouw, *Netten voor distributie van elektriciteit*. Arnhem: Phase to Phase, 2011 [Online]. Available: https://www.phasetophase.nl/boek/

[3] R. Lindsey, "Climate and Earth's Energy Budget," *NASA Earth Observatory*, 2009. [Online]. Available: https://earthobservatory.nasa.gov/features/EnergyBalance/page1.php

[4] A. Van Wijk, "Hydrogen key to a carbon-free energy system," in *Hydrogen Storage for Sustainability*, M. van de Voorde, Ed. Berlin, Boston: De Gruyter, 2021, pp. 43–104 [Online]. Available: https://www.degruyter.com/document/doi/10.1515/9783110596281-005/html

[5] V. Smil, *Power Density: a key to understanding energy sources and uses*. MIT Press, 2015.

[6] E. G. Hertwich *et al.*, "Integrated life-cycle assessment of electricity-supply scenarios confirms global environmental benefit of low-carbon technologies," *Proc. Natl. Acad. Sci. U. S. A.*, vol. 112, no. 20, pp. 6277–6282, 2015, doi: 10.1073/pnas.1312753111. [Online]. Available: http://www.scopus.com/inward/record.url?eid=2-s2.0-84929404255&partnerID=tZOtx3y1

[7] T. Smolinka, "Water Electrolysis: Status and Potential for Development," Freiburg, 2014.

[8] G. Glenk and S. Reichelstein, "Reversible Power-to-Gas systems for energy conversion and storage," *Nat. Commun.*, vol. 13, no. 2010, 2022, doi: 10.1038/s41467-022-29520-0. [Online]. Available: https://doi.org/10.1038/s41467-022-29520-0

[9] Solenco Power, "SOLENCO Powerbox", 2017 [Online]. Available: https://www.energystoragenl.nl/wp-content/uploads/2017/02/Solenco-Power_SPB-Brochure.pdf

[10] I. Jakupca, "NASA Fuel Cell and Hydrogen Activities," no. April, 2019 [Online]. Available: https://www.hydrogen.energy.gov/pdfs/review19/ia011_jakupca_2019_o.pdf

[11] S. Amelang and Sunfire, "Europe must take ambitious lead in green hydrogen – German govt advisor," 2020. [Online]. Available: https://www.cleanenergywire.org/news/europe-must-take-ambitious-lead-green-hydrogen-german-govt-advisor

[12] H. van 't Noordende and P. Ripson, "A One-GigaWatt Green-Hydrogen Plant - Advanced Design and total Installed-Capital Costs," 2022 [Online]. Available: https://ispt.eu/media/ISPT-public-report-gigawatt-green-hydrogen-plant.pdf

[13] A. J. M. van Wijk, "Hydrogen as key in the energy transition," in *Offshore renewable energy course*, 2023.

[14] E. Hand, "Hidden Hydrogen," *Science (80-.)*, vol. 379, no. 6633, pp. 630–636, 2023, doi: 10.1126/science.adh1477.

[15] IPCC, "Summary for policymakers," in *Climate Change 2021: The Physical Science Basis. Contribution of Working Group I to the Sixth Assessment Report of the Intergovernmental Panel on Climate Change*, V. Masson-Delmotte, P. Zhai, A. Pirani, S. L. Connors, C. Péan, S. Berger, N. Caud, Y. Chen, L. Goldfarb, M. I. Gomis, M. Huang, K. Leitzell, E. Lonnoy, J. B. R. Matthews, T. K. Maycock, T. Waterfield, O. Yelekçi, R. Yu, and B. Zhou, Eds. Cambridge, United Kingdom and New York, NY, USA: Cambridge University Press, 2021 [Online]. Available: https://www.ipcc.ch/report/ar6/wg1/

[16] A. Arrigoni and L. Bravo Diaz, *Hydrogen emissions from a hydrogen economy and their potential global warming impact: summary report of the Clean Hydrogen Joint Undertaking expert workshop on the Environmental Impacts of Hydrogen*. Publications Office of the European Union, 2022 [Online]. Available: https://publications.jrc.ec.europa.eu/repository/handle/JRC130362

[17] IRENA, *Renewable Power Generation Costs in 2020*. Abu Dhabi, 2020 [Online]. Available: https://www.irena.org/publications/2021/Jun/Renewable-Power-Costs-in-2020

[18] A. van Wijk and F. Wouters, "Hydrogen - The bridge between Africa and Europe," in *Shaping an Inclusive Energy Transition*, 1st ed., M. C. P. Weijnen, Z. Lukszo, and S. Farahani, Eds. Springer Cham, 2021, pp. 91–120 [Online]. Available: https://link.springer.com/chapter/10.1007/978-3-030-74586-8_5

[19] A. van Wijk and J. Chatzimarkakis, "Green Hydrogen for a European Green Deal - A 2x40 GW Initiative," 2020 [Online]. Available: https://static1.squarespace.com/static/5d3f0387728026000121b2a2/t/5e85aa53179bb450f86a4efb/1585818266517/2020-04-01_Dii_Hydrogen_Studie2020_v13_SP.pdf

[20] IEA, "The Future of Hydrogen," Paris, 2019 [Online]. Available: https://www.iea.org/publications/reports/thefutureofhydrogen/

[21] Hydrogen Council and McKinsey, "Hydrogen Insights, A perspective on hydrogen investment, market development and cost competitiveness," 2021 [Online]. Available: https://hydrogencouncil.com/wp-content/uploads/2021/02/Hydrogen-Insights-2021.pdf

[22] Bloomberg NEF, "Hydrogen Economy Outlook," 2020 [Online]. Available: https://data.bloomberglp.com/professional/sites/24/BNEF-Hydrogen-Economy-Outlook-Key-Messages-30-Mar-2020.pdf

[23] Leiden-Delft-Erasmus Centre for Sustainability Circular Industries Hub, "Critical materials, Green Energy and Geopolitics: A Complex mix," pp. 1–52, 2022 [Online]. Available: https://www.centre-for-sustainability.nl/uploads/cfs/attachments/Critical Materials_LDE White Paper_DEF20220627_0.pdf

[24] IEA, "The Role of Critical Minerals in Clean Energy Transitions," Paris, 2022 [Online]. Available: https://www.iea.org/reports/the-role-of-critical-minerals-in-clean-energy-transitions

[25] IEA, "Material efficiency in clean energy transitions," Paris, 2019 [Online]. Available: https://www.iea.org/reports/material-efficiency-in-clean-energy-transitions

[26] M. van der Star, "Material Requirements for Water Electrolysers in the Netherlands 2020-2050, The necessity of innovations in scaling up," Leiden-Delft Universities, 2022.

[27] World Bank, "Report: Universal Access to Sustainable Energy Will Remain Elusive Without Addressing Inequalities," 2021. [Online]. Available: https://www.worldbank.org/en/news/press-release/2021/06/07/report-universal-access-to-sustainable-energy-will-remain-elusive-without-addressing-inequalities

[28] M. P. C. Weijnen, Z. Lukszo, and S. Farahani, Eds., *Shaping an Inclusive Energy Transition*. Cham: Springer International Publishing, 2021 [Online]. Available: https://link.springer.com/10.1007/978-3-030-74586-8

[29] A. van Wijk, E. van der Roest, and J. Boere, *Solar power to the people (ENG)*. Amsterdam: IOS Press BV, 2017 [Online]. Available: https://www.alliedwaters.com/wp-content/uploads/2017/11/19-12-ENG-Solar-Power-to-the-people.pdf

[30] IEA, "Key World Energy Statistics 2021," 2021 [Online]. Available: https://www.iea.org/data-and-statistics/data-product/world-energy-balances

[31] Our World in Data, "Energy use per person, 2021," 2021. [Online]. Available: https://ourworldindata.org/grapher/per-capita-energy-use?tab=chart

[32] BP, "bp Statistical Review of World Energy 2021," 2022 [Online]. Available: https://www.bp.com/en/global/corporate/energy-economics/statistical-review-of-world-energy.html

[33] Solargis and World Bank Group, "Global Solar Atlas 2.0 - GHI - Global horizontal irradiation," 2022. [Online]. Available: https://globalsolaratlas.info/download/world

[34] DTU Wind Energy and World Bank Group/ESMAP, "Global Wind Atlas 3.0 - GIS files & API access," 2022. [Online]. Available: https://globalsolaratlas.info/download/world

[35] USGS, "New Map of Worldwide Croplands Supports Food and Water Security," 2017. [Online]. Available: https://www.usgs.gov/news/featured-story/new-map-worldwide-croplands-supports-food-and-water-security

[36] H. Ritchie and M. Roser, "Land Use," *Our World in Data*, 2019. [Online]. Available: https://ourworldindata.org/land-use

[37] P. van Son and T. Isenburg, *Emission free energy from the deserts*. The Hague: Smart Book Publishers, 2019.

[38] Center for International Earth Science Information Network - CIESIN - Columbia University, "Gridded Population of the World, Version 4 (GPWv4): Population Density, Revision 11." NASA Socioeconomic Data and Applications Center (SEDAC), Palisades, New York, 2018 [Online]. Available: https://doi.org/10.7927/H49C6VHW

[39] Our World in Data, "Population density, 2022," 2022. [Online]. Available: https://ourworldindata.org/grapher/population-density

[40] Esri Deutschland, "World airports," 2022. [Online]. Available: https://hub.arcgis.com/datasets/esri-de-content::world-airports/explore?location=26.691812%2C53.275203%2C6.69

[41] Solargis and World Bank Group, "Global Solar Atlas 2.0 - PVOUT," 2022. [Online]. Available: https://globalsolaratlas.info/download/world

[42] J. Latham, R. Cumani, I. Rosati, and M. Bloise, "FAO Global Land Cover (GLC-SHARE) Beta-Release 1.0 Database (layer 2,4,5,10,11)," 2014. [Online]. Available: https://data.apps.fao.org/map/catalog/srv/eng/catalog.search#/metadata/ba4526fd-cdbf-4028-a1bd-5a559c4bff38

[43] UNEP-WCMC and IUCN, "Protected Planet: The World Database on Protected Areas (WDPA)." UNEP-WCMC and IUCN, Cambridge, UK, 2021 [Online]. Available: https://www.protectedplanet.net/en

[44] F. H. Saadi, N. S. Lewis, and E. W. McFarland, "Relative costs of transporting electrical and chemical energy," *Energy Environ. Sci.*, vol. 11, no. 3, pp. 469–475, 2018, doi: 10.1039/c7ee01987d.

[45] E. Bellini, "Saudi Arabia's second PV tender draws world record low bid of $0.0104/kWh," *PV magazine*, 2021. [Online]. Available: https://www.pv-magazine.com/2021/04/08/saudi-arabias-second-pv-tender-draws-world-record-low-bid-of-0104-kwh/

[46] Wind Europe, "Onshore wind energy scores lowest ever price under new Spanish auction design," 2021. [Online]. Available: https://windeurope.org/newsroom/news/onshore-wind-energy-scores-lowest-ever-price-under-new-spanish-auction-design/

[47] IRENA, *Renewable Power Generation Costs in 2021*, Abu Dhabi: International Renewable Energy Agency, 2022 [Online]. Available: https://www.irena.org/-/media/Files/IRENA/Agency/Publication/2018/Jan/IRENA_2017_Power_Costs_2018.pdf

[48] Forest Research, "Typic caloric values of fuels." [Online]. Available: https://www.forestresearch.gov.uk/tools-and-resources/fthr/biomass-energy-resources/reference-biomass/facts-figures/typical-calorific-values-of-fuels/

[49] The Engineering Toolbox, "Fuels - Higher and Lower Calorific Values." [Online]. Available: https://www.engineeringtoolbox.com/fuels-higher-calorific-values-d_169.html

[50] The Engineering Toolbox, "Ammonia - Thermophysical Properties." [Online]. Available: https://www.engineeringtoolbox.com/ammonia-d_1413.html

[51] P. P. Lopes and V. R. Stamenkovic, "Past, present, and future of lead–acid batteries," *Science (80-.)*, vol. 369, no. 6506, pp. 923–924, 2020, doi: 10.1126/science.abd3352. [Online]. Available: https://doi.org/10.1126/science.abd3352

[52] A. Smets, K. Jäger, O. Isabella, R. van Swaai, and M. Zeman, "19.3 Batteries," in *Solar Energy*, Cambridge: UIT, 2016 [Online]. Available: https://ocw.tudelft.nl/wp-content/uploads/solar_energy_section_19_3_1-19_3_2.pdf

[53] D. DeSantis, B. D. James, C. Houchins, G. Saur, and M. Lyubovsky, "Cost of long-distance energy transmission by different carriers," *iScience*, vol. 24, no. 12, p. 103495, 2021, doi: 10.1016/j.isci.2021.103495. [Online]. Available: https://doi.org/10.1016/j.isci.2021.103495

[54] K. C. Kavvadias and S. Quoilin, "Exploiting waste heat potential by long distance heat transmission: Design considerations and techno-economic assessment," *Appl. Energy*, vol. 216, no. February, pp. 452–465, 2018, doi: 10.1016/j.apenergy.2018.02.080. [Online]. Available: https://doi.org/10.1016/j.apenergy.2018.02.080

[55] European Commission, "Gas storage - Ensuring shared gas storage capacities in the EU is crucial to guarantee security of energy supply," 2022. [Online]. Available: https://energy.ec.europa.eu/topics/energy-security/gas-storage_en

[56] J. Michalski, U. Bu, F. Crotogino, and S. Donadei, "Hydrogen generation by electrolysis and storage in salt caverns: Potentials , economics and systems aspects with regard to the German energy transition," *Int. J. Hyd*, vol. 42, no. 19, pp. 13427–13443, 2017, doi: 10.1016/j.ijhydene.2017.02.102.

[57] Energy.gov, "Strategic Petroleum Reserve (SPR) FAQs." [Online]. Available: https://www.energy.gov/ceser/spr-faqs#:~:text=Stockpiling oil in artificially-created, least five times the expense

[58] M. Aneke and M. Wang, "Energy storage technologies and real life applications – A state of the art review," *Appl. Energy*, vol. 179, no. July 2020, pp. 350–377, 2016, doi: 10.1016/j.apenergy.2016.06.097.

[59] R. Bollini, "Understanding self-discharge of a Lithium-ion battery," 2022. [Online]. Available: https://evreporter.com/understanding-self-discharge-of-a-lithium-ion-battery/

[60] H. Derking, L. van der Togt, and M. Keezer, "Liquid Hydrogen Storage: Status and Future Perspectives," *Cyrogenic Heat and Mass Transfer Congress*. Enschede, The Netherlands, 2019 [Online]. Available: https://documents.pub/document/liquid-hydrogen-storage-status-and-future-perspectives-all-liquid-hydrogen-storage.html?page=1

[61] M. Bijkerk, *Help de energietransitie*. Stili Novi, 2022.

[62] K. Mongird *et al.*, "Energy Storage Technology and Cost Characterization Report | Department of Energy," 2019 [Online]. Available: https://www.energy.gov/eere/water/downloads/energy-storage-technology-and-cost-characterization-report

[63] R. van der Pluijm, "HyStock, connecting and distributing electrons and molecules," in *PowerWeb Conference*, 2018 [Online]. Available: https://d2k0ddhflgrk1i.cloudfront.net/Websections/Powerweb/Annual Conference 2018/Robbert van der Pluijm Energy Stock juni 2018.pdf

[64] EnergiNet and T. D. E. Agency, "Technology Data Energy Storage," 2020 [Online]. Available: https://ens.dk/en/our-services/projections-and-models/technology-data/technology-data-energy-storage

[65] N. Hartog, M. Bloemendal, E. Slingerland, and A. van Wijk, "Duurzame warmte gaat ondergronds," Nieuwegein, 2016 [Online]. Available: https://library.kwrwater.nl/publication/54683679/

[66] A. Gupta *et al.*, "Hydrogen Clathrates: Next Generation Hydrogen Storage Materials," *Energy Storage Mater.*, vol. 41, no. March, pp. 69–107, 2021, doi: 10.1016/j.ensm.2021.05.044.

[67] J. G. Speight, "3 - Unconventional gas," 2nd ed., J. G. Speight, Ed. Boston: Gulf Professional Publishing, 2019, pp. 59–98 [Online]. Available: https://www.sciencedirect.com/science/article/pii/B9780128095706000035

[68] Wusel007/Wikimedia Commons, "Unlocking the mystery of methane clathrates — on Earth and on our solar system's icy moons," *Georgia Tech - School of Civil and Environmental Engineering*, 2019. [Online]. Available: https://ce.gatech.edu/news/unlocking-mystery-methane-clathrates-earth-and-our-solar-system-s-icy-moons

[69] P. Quintela De Saldanha, "Sines H2 Hub a cost perspective of the transmission & storage infrastructure of the Sines green hydrogen hub [MSc. Thesis]," Delft University of Technology, 2021 [Online]. Available: http://resolver.tudelft.nl/uuid:3d395140-7169-4e0f-8e75-ec3b2d8924dc

[70] Gas for Climate and Guidehouse, "Extending the European Hydrogen Backbone; A European Hydrogen Infrastructure vision covering 21 countries," 2021 [Online]. Available: https://www.ehb.eu/files/downloads/European-Hydrogen-Backbone-April-2021-V3.pdf

[71] Panasonic Group, "Panasonic Launches 5 kW Type Pure Hydrogen Fuel Cell Generator," 2021. [Online]. Available: https://news.panasonic.com/global/press/en211001-4

[72] Government of Dubai, "Mohammed bin Rashid Al Maktoum Solar Park." [Online]. Available: https://www.mbrsic.ae/en/about/mohammed-bin-rashid-al-maktoum-solar-park/

[73] Bloomberg News, "Giant China Project Leads the Rise of Renewable Mega-Hubs," 2021. [Online]. Available: https://www.bloomberg.com/news/articles/2021-10-13/xi-s-solar-wind-hub-gives-china-a-lead-in-supersize-renewables#xj4y7vzkg

[74] I. Roger, M. A. Shipman, and M. D. Symes, "Earth-abundant catalysts for electrochemical and photoelectrochemical water splitting," *Nat. Rev. Chem.*, vol. 1, no. 1, p. 3, 2017, doi: 10.1038/s41570-016-0003. [Online]. Available: https://doi.org/10.1038/s41570-016-0003

[75] M. Maisch, "[PV Magazine] Hydrogen-producing rooftop solar panels nearing commercialization," 2022. [Online]. Available: https://solhyd.org/nl/solhyd-in-de-media/pv-magazine-hydrogen-producing-rooftop-solar-panels-nearing-commercialization/

[76] Y. Zhou and R. S. J. Tol, "Evaluating the costs of desalination and water transport," *Water Resour. Res.*, vol. 41, no. 3, pp. 1–10, Mar. 2005, doi: 10.1029/2004WR003749. [Online]. Available: http://doi.wiley.com/10.1029/2004WR003749

[77] A. Aende, J. Gardy, and A. Hassanpour, "Seawater desalination: A review of forward osmosis technique, its challenges, and future prospects," *Processes*, vol. 8, no. 8, 2020, doi: 10.3390/PR8080901.

[78] J. Kim, K. Park, D. R. Yang, and S. Hong, "A comprehensive review of energy consumption of seawater reverse osmosis desalination plants," *Appl. Energy*, vol. 254, no. August, p. 113652, Nov. 2019, doi: 10.1016/j.apenergy. 2019.113652. [Online]. Available: https://linkinghub.elsevier.com/retrieve/pii/S030626191931339X

[79] NASA Aquarius, "Constituents of Seawater," 2018. [Online]. Available: https://salinity.oceansciences.org/learn-more.htm?id=54

[80] H.-W. Balling, M. Janse, and P. Sondervan, "Trace elements, functions, sinks and replenishment in reef aquaria," in *Advances in coral husbandry in public aquariums,* Burgers' Zoo, Arnhem, The Netherlands, 2008, pp. 143–156.

[81] M. van Baarlen *et al.*, "Hydrogen production combined with brine processing in the desert." ChemE - Faculty of Applied Sciences, Delft University of Technology (student group), Delft, 2020.

[82] "Neom." [Online]. Available: https://www.neom.com/en-us

[83] M. Goff, "Distribution of Ammonia as an Energy Carrier." 2020 [Online]. Available: https://www.ammoniaenergy.org/wp-content/uploads/2020/12/Michael-Goff.pdf

[84] NuStar, "Pipeline Transportation of Ammonia." 2021 [Online]. Available: https://www.ammoniaenergy.org/wp-content/uploads/2021/11/AEA-Ammonia-Pipeline-Transportation-MEA-11-4-2021.pdf

[85] Academic Gain Tutorials, "Solar Vapor Absorption Refrigeration System (Ammonia-Water Solar Cooling System) Explained," 2020. [Online]. Available: https://www.youtube.com/watch?v=PjcdqAkP0UA

[86] M. Ivanova, "Top 10 offshore producing countries in 2017," 2017. [Online]. Available: https://www.offshore-mag.com/production/article/16756124/top-10-offshore-producing-countries-in-2017

[87] IRENA, "Floating Foundations: A Game Changer for Offshore Wind Power," Abu Dhabi, 2016 [Online]. Available: http://www.irena.org/-/media/Files/IRENA/Agency/Publication/2016/IRENA_Offshore_Wind_Floating_Foundations_2016.pdf

[88] BBC, "Floating wind farm records UK's top results for potential output again," 2021. [Online]. Available: https://www.bbc.com/news/uk-scotland-north-east-orkney-shetland-56496355

[89] A. Viré, "Floating wind turbines change everything," 2022. [Online]. Available: https://www.tudelft.nl/en/2022/tu-delft/floating-wind-turbines-change-everything

[90] Boskalis, "Boskalis: Kincardine is a prelude to more floating wind projects," 2021. [Online]. Available: https://www.projectcargojournal.com/offshore/2021/02/18/boskalis-kincardine-is-a-prelude-to-more-floating-wind-projects/?gdpr=accept&gdpr=deny

[91] ERM, D. Caine, M. Iliffe, K. Kinsella, W. Wahyuni, and L. Bond, "Dolphyn Hydrogen Phase 1 - Final Report," *UK Department for Business, Energy and Industrial Strategy*. 2019 [Online]. Available: https://www.gov.uk/government/publications/hydrogen-supply-competition

[92] Ampelmann, "Offshore wind." [Online]. Available: https://www.ampelmann.nl/offshore-wind

[93] J. Juez-Larré, S. van Gessel, R. Dalman, G. Remmelts, and R. Groenenberg, "Assessment of underground energy storage potential to support the energy transition in the Netherlands," *First Break*, vol. 37, no. 7, pp. 57–66, 2019.

[94] Eurostat, "Energy statistics - an overview," 2022. [Online]. Available: https://ec.europa.eu/eurostat/statistics-explained/index.php?title=Energy_statistics_-_an_overview#Final_energy_consumption

[95] "OCEANERGY." [Online]. Available: http://www.oceanergy.com/

[96] OCEANERGY, "Kite Hydrogen Ships." [Online]. Available: https://www.oceanergy.com/kite-gas-fuel-ship/

[97] Smartland, "Amazon investeert 1,5 miljoen euro in North Sea Farm 1 project," 2023. [Online]. Available: https://windenergie-nieuws.nl/16/amazon-investeert-1-5-miljoen-euro-in-north-sea-farm-1-project/

[98] Siemens Gamesa, "The SG 14-222 DD." [Online]. Available: https://www.siemensgamesa.com/-/media/siemensgamesa/downloads/en/products-and-services/offshore/brochures/siemens-gamesa-offshore-wind-turbine-brouchure-sg-14-222-dd.pdf?ste_sid=1896e116336aa964dbce863e72fa4386

[99] Stichting Noordzeeboerderij, "Sea the Power of Sustainable Farming." 2020 [Online]. Available: https://bio4safe.eu/sites/default/files/2021-03/PUBL_IR1601_RoadmapSeaweedBiostimulants_0.pdf

[100] World Bank, "Seaweed Aquaculture for Food Security, Income Generation and Environmental Health in Tropical Developing Countries." 2016 [Online]. Available: https://documents1.worldbank.org/curated/en/947831469090666344/pdf/107147-WP-REVISED-Seaweed-Aquaculture-Web.pdf

[101] N. Muradov, "Low-carbon production of hydrogen from fossil fuels," in *Compendium of Hydrogen Energy*, V. Subramani, A. Basile, and T. N. B. T.-C. of H. E. Veziroğlu, Eds. Oxford: Woodhead Publishing, 2015, pp. 489–522 [Online]. Available: https://www.sciencedirect.com/science/article/pii/B9781782423614000170

[102] P. Webb, "Overview of the oceans," in *Introduction to Oceanography*, 2019 [Online]. Available: https://rwu.pressbooks.pub/webboceanography/chapter/1-1-overview-of-the-oceans/

[103] Z. Liu, Z. Deng, S. J. Davis, C. Giron, and P. Ciais, "Monitoring global carbon emissions in 2021," *Nat. Rev. Earth Environ.*, vol. 3, no. 4, pp. 217–219, 2022, doi: 10.1038/s43017-022-00285-w.

[104] A. Frangoul, “A pilot project in the North Sea will develop floating solar panels that glide over waves ‘like a carpet,’” *CNBC*, 2022. [Online]. Available: https://www.cnbc.com/2022/07/22/europes-energy-giants-explore-potential-of-floating-solar-.html

[105] AFSIA, “JinkoSolar Signs Strategic Cooperation Agreement with CIMC Raffles for Offshore Solar Projects,” 2022. [Online]. Available: http://afsiasolar.com/jinkosolar-signs-strategic-cooperation-agreement-with-cimc-raffles-for-off-shore-solar-projects/

[106] Oceans of Energy, “Oceans of Energy - North Sea 3,” 2021. [Online]. Available: https://oceansofenergy.blue/2021/09/01/eu-scores-project-aims-to-deliver-world-first-bankable-hybrid-offshore-marine-energy-parks/

[107] Ocean Sun, “The Institute for Energy Technology quantifies Ocean Sun’s performance gain,” 2020. [Online]. Available: https://oceansun.no/the-institute-for-energy-technology-quantifies-ocean-suns-performance-gain/

[108] C. Zervas, “Offshore Solar to Hydrogen System, Located North of Crete, Greece: A techno-economic assessment [MSc. Thesis],” Delft University of Technology, 2021 [Online]. Available: Offshore solar to hydrogen

[109] Ecopower International, “How does it work.” [Online]. Available: http://www.ecopowerinternational.com/html/how_does_it_work.html

[110] C. A. Infante Ferreira, “Thermal Gradient (OTEC).” [Online]. Available: https://www.tudelft.nl/oceanenergy/research/thermal-gradient-otec

[111] R. Kleiterp, “The feasibility of a commercial osmotic power plant [MSc. thesis],” Delft University of Technology, 2012 [Online]. Available: http://resolver.tudelft.nl/uuid:fbaa8d2f-3c01-45e3-8473-9a2ccd2b9a67

[112] W. Spoor, “Fotosynthese & Verbranding.” [Online]. Available: https://www.meneerspoor.nl/fotosyntheseverbranding.html

[113] X.-G. Zhu, S. P. Long, and D. R. Ort, “What is the maximum efficiency with which photosynthesis can convert solar energy into biomass?,” *Curr. Opin. Biotechnol.*, vol. 19, no. 2, pp. 153–159, 2008, doi: https://doi.org/10.1016/j.copbio.2008.02.004.

[114] Ministerie van Economische Zaken en Klimaat, “Contouren Nationaal plan energiesysteem.” Rijksoverheid, 2022 [Online]. Available: https://www.rijksoverheid.nl/documenten/kamerstukken/2020/07/01/kamerbrief-uitvoering-motie-gebruik-vaste-houtige-biomassa-voor-energietoepassingen

[115] D. Knöbel, “Renewed perspective on the role of biogas as local and or regional source of bio-hydrogen and bio-carbon dioxide within the future renewable hydrogen energy system [MSc. Thesis],” Delft University of Technology, 2022 [Online]. Available: http://resolver.tudelft.nl/uuid:6f0eae47-fc05-4e72-aedd-4f4d6907b44e

[116] P. Singh, C. Carliell-Marquet, and A. Kansal, “Energy pattern analysis of a wastewater treatment plant,” *Appl. Water Sci.*, vol. 2, no. 3, pp. 221–226, 2012, doi: 10.1007/s13201-012-0040-7.

[117] M. Maktabifard, E. Zaborowska, and J. Makinia, *Achieving energy neutrality in wastewater treatment plants through energy savings and enhancing renewable energy production*, vol. 17, no. 4. Springer Netherlands, 2018 [Online]. Available: https://doi.org/10.1007/s11157-018-9478-x

[118] EEA, “Water use in Europe — Quantity and quality face big challenges,” 2018. [Online]. Available: https://www.eea.europa.eu/signals/signals-2018-content-list/articles/water-use-in-europe-2014

[119] Eurostat, “Electricity and heat statistics,” 2022. [Online]. Available: https://ec.europa.eu/eurostat/statistics-explained/index.php?title=Electricity_and_heat_statistics#Consumption_of_electricity_per_capita_in_the_households_sector

[120] M. Gandiglio, A. Lanzini, A. Soto, P. Leone, and M. Santarelli, “Enhancing the Energy Efficiency of Wastewater Treatment Plants through Co-digestion and Fuel Cell Systems,” *Frontiers in Environmental Science*, vol. 5, 2017 [Online]. Available: https://www.frontiersin.org/articles/10.3389/fenvs.2017.00070

[121] CBS, “Zuivering van stedelijk afvalwater; procesgegevens afvalwater-behandeling,” 2022. [Online]. Available: https://www.cbs.nl/nl-nl/cijfers/detail/83399NED#StikstofverbindingenAlsN_5

[122] STOWA, “Emissie broeikasgassen vanuit RWZI’s,” 2012 [Online]. Available: https://www.stowa.nl/sites/default/files/assets/PUBLICATIES/Publicaties 2012/STOWA 2012-20 %28revisie 2019%29.pdf

[123] Vewin, “Drinkwaterstatistieken 2022,” 2022 [Online]. Available: https://www.vewin.nl/SiteCollectionDocuments/Publicaties/Cijfers/Vewin-Drinkwater-statistieken-2022-NL-WEB.pdf

[124] T. Flameling, L. Nieukoop, R. van der Lans, B. Reitsma, and A. Visser, “Waterstof-productie in RWZI’s met fijne-bellenbeluchtingssystemen.” STOWA, Amersfoort, 2022 [Online]. Available: https://www.stowa.nl/sites/default/files/assets/PUBLICATIES/Publicaties 2022/STOWA 2022-51 SEH.pdf

[125] E. van der Roest, R. Bol, T. Fens, and A. van Wijk, “Utilisation of waste heat from PEM electrolysers – Unlocking local optimisation,” *Int. J. Hydrogen Energy*, vol. in press, 2023.

[126] L. De Donno Novelli, S. Moreno Sayavedra, and E. R. Rene, “Polyhydroxy-alkanoate (PHA) production via resource recovery from industrial waste streams: A review of techniques and perspectives,” *Bioresour. Technol.*, vol. 331, p. 124985, 2021, doi: https://doi.org/10.1016/j.biortech.2021.124985. [Online]. Available: https://www.sciencedirect.com/science/article/pii/S0960852421003242

[127] M. Mehrpouya, H. Vahabi, M. Barletta, P. Laheurte, and V. Langlois, “Additive manufacturing of polyhydroxyalkanoates (PHAs) biopolymers: Materials, printing techniques, and applications,” *Mater. Sci. Eng. C*, vol. 127, p. 112216, 2021, doi: https://doi.org/10.1016/j.msec.2021.112216. [Online]. Available: https://www.sciencedirect.com/science/article/pii/S0928493121003568

[128] A. Giubilini, F. Bondioli, M. Messori, G. Nyström, and G. Siqueira, “Advantages of Additive Manufacturing for Biomedical Applications of Polyhydroxyalkanoates,” *Bioengineering*, vol. 8, no. 2, 2021.

[129] S. Matassa, D. J. Batstone, T. Hülsen, J. Schnoor, and W. Verstraete, “Can Direct Conversion of Used Nitrogen to New Feed and Protein Help Feed the World?,” *Environ. Sci. Technol.*, vol. 49, no. 9, pp. 5247–5254, May 2015, doi: 10.1021/es505432w. [Online]. Available: https://pubs.acs.org/doi/10.1021/es505432w

[130] “Power-to-Protein.” [Online]. Available: www.powertoprotein.eu

[131] F. Oesterholt, E. Broeders, and C. Zamalloa, “Power-to-Protein: eiwitproductie in een circulaire economie.” Nieuwegein, KWR 2018.078, 2019 [Online]. Available: http://api.kwrwater.nl//uploads/2019/03/KWR-2018.078-Power-to-protein-eiwitproductie-in-een-circulaire-economie.-Fase-2-Pilotonderzoek.pdf

[132] I. Tobío-Pérez, A. Alfonso-Cardero, Y. Díaz-Domínguez, S. Pohl, R. Piloto-Rodríguez, and M. Lapuerta, “Thermochemical Conversion of Sargassum for Energy Production: a Comprehensive Review,” *BioEnergy Res.*, no. 0123456789, 2022, doi: 10.1007/s12155-021-10382-1. [Online]. Available: https://doi.org/10.1007/s12155-021-10382-1

[133] The Pew Charitable Trusts, “Seaweed Farming Has Vast Potential (But Good Luck Getting a Permit),” 2022. [Online]. Available: https://www.pewtrusts.org/en/research-and-analysis/blogs/stateline/2022/03/07/seaweed-farming-has-vast-potential-but-good-luck-getting-a-permit

[134] Scubazoo blog, “Food, fuel and more will be produced in sea farms of future,” 2014. [Online]. Available: https://innovationtoronto.com/2014/10/food-fuel-and-more-will-be-produced-in-sea-farms-of-future/

[135] J. Cai, “Global status of seaweed production, trade and utilization,” in *Seaweed Innovation Forum Belize*, 2021 [Online]. Available: https://www.competecaribbean.org/wp-content/uploads/2021/05/Global-status-of-seaweed-production-trade-and-utilization-Junning-Cai-FAO.pdf

[136] M. Wilde, “The Dead Zone,” 2019. [Online]. Available: https://www.sdgnederland.nl/nieuws/the-dead-zone/

[137] C. M. Duarte, J. Wu, X. Xiao, A. Bruhn, and D. Krause-Jensen, “Can Seaweed Farming Play a Role in Climate Change Mitigation and Adaptation?,” *Front. Mar. Sci.*, vol. 4, Apr. 2017, doi: 10.3389/fmars.2017.00100. [Online]. Available: http://journal.frontiersin.org/article/10.3389/fmars.2017.00100/full

[138] United Nations Environment Programme and Caribbean Environment Programme, “Sargassum White Paper: Turning the Crisis into an Opportunity.” 2021 [Online]. Available: https://wedocs.unep.org/20.500.11822/36244

[139] A. Desrochers, S.-A. Cox, H. A. Oxenford, and B. van Tussenbroek, “Sargassum Uses Guide: a resource for Caribbean researchers, entrepreneurs and policy makers,” 2020 [Online]. Available: https://sargassumhub.org/sargassum-uses-guide-now-available/

[140] J. Scully, “Ocean Sun calls for faster floating PV permitting, eyes utility-scale projects,” 2022. [Online]. Available: https://www.pv-tech.org/ocean-sun-calls-for-faster-floating-pv-permitting-eyes-utility-scale-projects/

[141] FAO, *Report of the Third Annual Project Steering Committee Meeting – Climate Change Adaptation in the Eastern Caribbean Fisheries Sector (CC4FISH)*, no. 1332. Bridgetown, Barbados, 16-17 april 2019, 2021 [Online]. Available: https://www.fao.org/publications/card/fr/c/CB3817EN/

[142] E. van der Roest, T. Fens, M. Bloemendal, S. Beernink, J. P. van der Hoek, and A. J. M. van Wijk, “The Impact of System Integration on System Costs of a Neighborhood Energy and Water System,” *Energies*, vol. 14, no. 9, p. 2616, May 2021, doi: 10.3390/en14092616. [Online]. Available: https://www.mdpi.com/1996-1073/14/9/2616

[143] DVGW and Hydrogen Europe, “Pyrolysis - Potential and possible applications of a climate-friendly hydrogen production,” 2022 [Online]. Available: https://hydrogeneurope.eu/wp-content/uploads/2022/10/ewp_kompakt_pyrolyse_english_web.pdf

[144] A. Stroo, “Offshore Methane Pyrolysis: A techno-economic analysis to assess the feasibility of offshore methane pyrolysis for the production of hydrogen [MSc Thesis],” Delft University of Technology, 2023 [Online]. Available: http://resolver.tudelft.nl/uuid:662993e5-2ff1-451d-a7d5-3a31fb4c722f

[145] Monolith, “Monolith - Photos and Videos.” [Online]. Available: https://monolith-corp.com/photos-and-videos

[146] F. Wouters and A. van Wijk, “Speed, the forgotten cost reduction factor in the energy transition,” *Illuminem*, 2022. [Online]. Available: https://illuminem.com/illuminemvoices/speed-the-forgotten-cost-reduction-factor-in-the-energy-transition

[147] Z. Hausfather, “Explainer: How ‘Shared Socioeconomic Pathways’ explore future climate change,” 2018. [Online]. Available: https://www.carbonbrief.org/explainer-how-shared-socioeconomic-pathways-explore-future-climate-change/

[148] K. Riahi *et al.*, “The Shared Socioeconomic Pathways and their energy, land use, and greenhouse gas emissions implications: An overview,” *Glob. Environ. Chang.*, vol. 42, pp. 153–168, Jan. 2017, doi: 10.1016/j.gloenvcha.2016.05.009.

[149] S. KC and W. Lutz, “The human core of the shared socioeconomic pathways: Population scenarios by age, sex and level of education for all countries to 2100,” *Glob. Environ. Chang.*, vol. 42, pp. 181–192, Jan. 2017, doi: 10.1016/j.gloenvcha.2014.06.004.

[150] R. Dellink, J. Chateau, E. Lanzi, and B. Magné, “Long-term economic growth projections in the Shared Socioeconomic Pathways,” *Glob. Environ. Chang.*, vol. 42, pp. 200–214, Jan. 2017, doi: 10.1016/j.gloenvcha.2015.06.004.

[151] J. Crespo Cuaresma, “Income projections for climate change research: A framework based on human capital dynamics,” *Glob. Environ. Chang.*, vol. 42, pp. 226–236, Jan. 2017, doi: 10.1016/j.gloenvcha.2015.02.012.

[152] M. Leimbach, E. Kriegler, N. Roming, and J. Schwanitz, “Future growth patterns of world regions – A GDP scenario approach,” *Glob. Environ. Chang.*, vol. 42, pp. 215–225, Jan. 2017, doi: 10.1016/j.gloenvcha.2015.02.005.

[153] L. Jiang and B. C. O’Neill, “Global urbanization projections for the Shared Socioeconomic Pathways,” *Glob. Environ. Chang.*, vol. 42, pp. 193–199, Jan. 2017, doi: 10.1016/j.gloenvcha.2015.03.008.

[154] N. Bauer *et al.*, “Shared Socio-Economic Pathways of the Energy Sector – Quantifying the Narratives,” *Glob. Environ. Chang.*, vol. 42, pp. 316–330, Jan. 2017, doi: 10.1016/j.gloenvcha.2016.07.006.

[155] IPCC, *Climate Change 2022, Mitigation of Climate Change Summary for Policymakers (SPM)*. Cambridge, UK and New York, USA: Cambridge University Press, 2022 [Online]. Available: https://www.ipcc.ch/report/ar6/wg3/

[156] O. Fricko *et al.*, “The marker quantification of the Shared Socioeconomic Pathway 2: A middle-of-the-road scenario for the 21st century,” *Glob. Environ. Chang.*, vol. 42, pp. 251–267, Jan. 2017, doi: 10.1016/j.gloenvcha.2016.06.004.

[157] C. Berg and F. Kähler, “Turning off the Tap for Fossil Carbon Future Prospects for a Global Chemical and Derived Authors,” no. April, 2021 [Online]. Available: https://renewable-carbon.eu/publications/product/turning-off-the-tap-for-fossil-carbon-future-prospects-for-a-global-chemical-and-derived-material-sector-based-on-renewable-carbon/

[158] J. Gao, “Global 1-km Downscaled Population Base Year and Projection Grids Based on the Shared Socioeconomic Pathways, Revision 01.” NASA Socioeconomic Data and Applications Center (SEDAC), Palisades, New York, 2020 [Online]. Available: https://doi.org/10.7927/q7z9-9r69

[159] D. Eradus, “The Techno-Economic Feasibility of Green Hydrogen Storage in Salt Caverns in the Dutch North Sea [MSc. Thesis],” Delft University of Technology, 2022 [Online]. Available: http://resolver.tudelft.nl/uuid:8eb96cf8-2c91-4553-b0cb-a41458f61b5d

[160] S. Timmerberg and M. Kaltschmitt, “Hydrogen from renewables: Supply from North Africa to Central Europe as blend in existing pipelines – Potentials and costs,” *Appl. Energy*, vol. 237, no. July 2018, pp. 795–809, 2019, doi: 10.1016/j.apenergy.2019.01.030. [Online]. Available: https://doi.org/10.1016/j.apenergy.2019.01.030

[161] Hydrogen Council, “How hydrogen empowers the energy transition,” 2017 [Online]. Available: http://hydrogencouncil.com/wp-content/uploads/2017/11/Hydrogen-scaling-up-Hydrogen-Council.pdf

[162] V. D. W. M. Oldenbroek, “Integrated transport and energy systems based on hydrogen and fuel cell electric vehicles,” Delft University of Technology, 2021 [Online]. Available: https://doi.org/10.4233/uuid:f8f6566e-e50a-47e2-b1f9-67503ca1d021

[163] Z. Deng, S. Mol, and J. P. Van Der Hoek, “Shower heat exchanger: Reuse of energy from heated drinking water for CO_2 reduction,” *Drink. Water Eng. Sci.*, vol. 9, no. 1, pp. 1–8, 2016, doi: 10.5194/dwes-9-1-2016.

[164] D. Bouziotas, D. van Duuren, H. J. van Alphen, J. Frijns, D. Nikolopoulos, and C. Makropoulos, “Towards circular water neighborhoods: Simulation-based decision support for integrated decentralized urban water systems,” *Water (Switzerland)*, vol. 11, no. 6, 2019, doi: 10.3390/w11061227.

[165] World Bank, “Water in agriculture,” 2022. [Online]. Available: https://www.worldbank.org/en/topic/water-in-agriculture#1

[166] L. Rayner and M. Fessler, “Fair Energy Transition For All - How to get there? EU Wide Recommendations,” 2022 [Online]. Available: https://epc.eu/content/PDF/2022/FETA_final_report.pdf

Photo credits

p. 4: Image montage Rolf Rosing, Studio Quellijn
p. 22: Sunfire GmbH
p. 35: Junior Kannah / AFP / ANP
p. 45: SolarCleano
p. 62: Wikimedia Commons
p. 70: DEWA
p. 73: Solhyd
p. 82: NASA
p. 86: Lyfted Media for Dominion Energy
p. 87: Joshua Bauer / NREL
p. 88: Lyfted Media for Dominion Energy
p. 92: David Wingate / Oceanergy
p. 93: Stefan Payne-Wardenaar / Smartland landscape architecture
p. 96: Børge Bjørneklett / Ocean Sun
p. 101: Colin Keldie / EMEC
p. 113: MDPI
p. 119: Robert F. Bukaty / AP Photo / ANP
p. 133: Monolith
p. 135: Hyundai
p. 147: Airbus
All other photos: iStockphoto / Getty Images

Biography

Ad van Wijk
Ad van Wijk (1956) is Emeritus Professor of Future Energy Systems at Delft University of Technology and Guest Professor at KWR Water Research Institute. He also has a range of advisory positions in companies and other organizations.

Els van der Roest
Els van der Roest (1991) is a researcher at KWR Water Research Institute specializing in sustainable concepts involving Water & Energy.
In addition, she is working on a PhD at Delft University of Technology on integrated renewable energy and water systems for neighborhoods.

Jos Boere
Jos Boere (1959) is the director van Allied Waters, which focuses on the implementation of innovative and sustainable concepts in a circular economy. Jos is also a joint founder of Hysolar, producer and supplier of green hydrogen.

Acknowledgements

We published *Solar Power to the People* in November 2017. It helped to put hydrogen as a clean energy carrier on the agenda of board tables and politicians. Six years have now passed. Several countries, as well as the European Union, have now formulated a hydrogen strategy. And many companies, organizations, and research and educational institutions are now active in the field of hydrogen. Yet there are still a lot of unknowns when it comes to hydrogen production, transportation, storage and use. And also with respect to the technology, market development, safety, environmental impact and so on. But above all, there is still a lack of familiarity and understanding when it comes to the function of hydrogen in a sustainable energy system.

So there was widespread demand for a book describing how we can develop a sustainable energy system, what goals it must meet and the role hydrogen will play in that system. We were happy to take up that challenge but we couldn't have done it without the support of a range of organizations, companies, research institutes and, most importantly, individuals.

First of all, we would like to thank **KWR Water Research Institute** for its support, both practical and financial. In particular, we would like to thank: **Diederik van Hasselt** † for the research, **Frank Oesterholt** for his contribution to the content, and **Steven Ros** and **Bernard Raterman** for the maps in GIS. And of course, our thanks go to the co-initiators **Hysolar, Hydrogen Europe** and **DII** for their support.

Important substantive contributions were made by Masters students from Delft University of Technology, who completed their graduate research under the supervision of Ad van Wijk. The results of six graduate studies and those of a student research group have found their way into this book. They are the work of **Martijn van der Star** (material use in electrolyzers), **Christos Zervas** (offshore solar hydrogen production), **Diaz Knöbel** (biomass waste stream conversion to H_2 and CO_2), **Aike Stroo** (hydrogen production with methane pyrolysis), **Pedro Quintela de Saldanha** (hydrogen storage at solar hydrogen production), **Deirdre Eradus** (hydrogen storage at offshore wind-hydrogen production) and the student group **Matthijs van Baarlen, Rosemary Bolt, Arturo Cerdá Berme, Tim Fennis, Thomas van Foeken, Jeannine Frijns** en **Meet Somaiya** (brine mining).

The review of the book improved the quality of the content. We wish to thank the reviewers **Professor Kornelis Blok** (Energy System Analysis, Delft University of Technology, Lead author IPCC 3rd, 4th and 6th Assessment Reports), **Dr. Noé van Hulst** (Vice Chair IPHE, Special Advisor Hydrogen IEA, Senior Fellow Clingendael International Energy Programme), **Drs. Kees van der Leun** (Director Common Futures), **Professor Zofia Lukszo** (Smart Energy Systems, Delft University of Technology), **Professor Peter Luscuere** (Building Physics & Services, Delft University of Technology, Guest Professor Tianjin University China), **Ir. Paul van Son** (President Dii Desert Energy, Honorary President European Federation of Energy Traders [EFET]) and **Ir. Frank Wouters** (Senior Vice President New Energy at Reliance Industries, Chairman MENA Hydrogen Alliance, Co-President Long Duration Energy Storage Council, Fellow Payne Institute – Colorado School of Mines).

Finally, many thanks to all the brilliant, courageous and kind people who are doing their part to make a sustainable energy system a reality for all.

Ad van Wijk, Els van der Roest, Jos Boere

Colophon
This book was created and produced by Five Fountains B.V. on request of the initiators (KWR Water Research Institute, Hysolar, Hydrogen Europe, Dii Desert Energy).

Authors: Ad van Wijk, Els van der Roest, Jos Boere
Publisher: Allied Waters B.V.
Translation: Jim Adams, France
Project management: Henk Leenaers, Lijn43, Utrecht
Design: Ontwerpstudio Spanjaard, Maartensdijk
Infographics: Michel van Elk, Tiel
Cartography: Annemieke Altena, Buitenpost
Text editing: Lijn43, Utrecht
Visual editing: Rolf Rosing, Amsterdam

ISBN: 9789082763737
NUR: 950
DOI: 10.5281/zenodo.8207489

www.greenenergyforall.nl
www.greenenergyforall.eu